The Analysis of
Time Series

Further information on the complete range of Chapman and Hall
statistics books is available from the publishers.

The Analysis of Time Series
An Introduction

Fourth edition

C. Chatfield

Reader in Statistics
The University of Bath

CHAPMAN AND HALL

LONDON • NEW YORK • TOKYO • MELBOURNE • MADRAS

UK	Chapman and Hall, 2–6 Boundary Row, London SE1 8HN
USA	Chapman and Hall, 29 West 35th Street, New York NY10001
JAPAN	Chapman and Hall Japan, Thomson Publishing Japan, Hirakawacho Nemoto Building, 7F, 1–7–11 Hirakawa-cho, Chiyoda-ku, Tokyo 102
AUSTRALIA	Chapman and Hall Australia, Thomas Nelson Australia, 102 Dodds Street, South Melbourne, Victoria 3205
INDIA	Chapman and Hall India, R. Seshadri, 32 Second Main Road, CIT East, Madras 600 035

First edition 1975
Second edition 1980
Third edition 1984
Reprinted 1985, 1987
Fourth edition 1989
Reprinted 1991

© 1975, 1980, 1984, 1989 C. Chatfield

Typeset in 10/12 Times by Cotswold Typesetting Ltd, Gloucester

Printed in Great Britain
by St Edmundsbury Press Ltd, Bury St Edmunds, Suffolk

ISBN 0–412–31820–2

British Library Cataloguing in Publication Data

Chatfield, Christopher.
　　The analysis of time series. – 4th ed.
　　1. Time series. Analysis
　　I. Title
　　519.5′5

　　ISBN 0–412–31820–2

Library of Congress Cataloging-in-Publication Data

Chatfield, Christopher.
　　The analysis of time series: an introduction/
　　C. Chatfield.—4th ed.
　　p. cm.
　　Bibliography: p.
　　Includes index.
　　ISBN 0–412–31820–2.
　　1. Time-series analysis. I. Title.
　　QA280.C4 1989
　　519.5′5—dc20   89–7142
　　　　　　　　　　　　　　　　　　　　　　　　CIP

Contents

To Liz

Alice sighed wearily. 'I think you might do something better with the time,' she said, 'than waste it in asking riddles that have no answers.'

'If you knew Time as well as I do,' said the Hatter, 'you wouldn't talk about wasting **it**. It's **him**.'

'I don't know what you mean,' said Alice.

'Of course you don't!' the Hatter said, tossing his head contemptuously. 'I dare say you never even spoke to Time!'

'Perhaps not,' Alice cautiously replied: 'but I know I have to beat time when I learn music.'

'Ah! that accounts for it,' said the Hatter. 'He won't stand beating.'

Lewis Carroll, Alice's Adventures in Wonderland

Preface to fourth edition

The analysis of time series can be a rather difficult topic, and my aim in writing this book has been to provide a comprehensible introduction which considers both theory and practice. Enough theory is given to introduce the concepts of time-series analysis and make the book mathematically interesting. In addition various practical examples are considered so as to help the reader tackle the analysis of real data.

The book can be used as a text for an undergraduate or postgraduate course in time series, or it can be used for self-tuition by research workers. The book assumes a knowledge of basic probability theory and elementary statistical inference. In order to keep the level of mathematics required as low as possible, some mathematical difficulties have been glossed over, particularly in Sections 3.4.8 and 3.5 which the reader is advised to omit at a first reading. Nevertheless a fair level of mathematical sophistication is required in places, as for example in the need to handle Fourier transforms, although I have helped the reader here by providing a special appendix on this topic

Although the book is primarily an introductory text, I have nevertheless added appropriate references to further reading and to more advanced topics so that the interested reader can pursue his studies if he wishes. These references are mainly to comprehensible and readily accessible sources rather than to the original attributive references.

One difficulty in writing an introductory textbook is that many practical problems contain at least one feature which is 'non-standard', and these cannot all be envisaged in a book of reasonable length. Thus the reader who has grasped the basic concepts of time-series analysis should always be prepared to use his common sense in tackling a problem. Example 5.1 is a typical situation where common sense has to be applied and also stresses the fact that the first step in any time-series analysis should be to plot the data. The worked examples in Appendix D also include candid comments on practical difficulties to complement the main text.

The first nine chapters of the fourth edition have much the same structure as the first three editions, but many sections have been largely rewritten, particularly in Chapters 2 to 5, and the text has been clarified and updated throughout. There is a new Chapter 10 on state-space models, as this topic is of much current interest. The old Chapter 10 has been extensively revised as a new Chapter 11. The references have been updated throughout.

I am indebted to many people for assistance in the preparation of this book, including V. Barnett, D. R. Cox, K. V. Diprose, R. Fenton, P. R. Fisk, H. Neave, J. Marshall, P. Newbold, M. P. G. Pepper, M. B. Priestley, D. L. Prothero, A. Robinson, B. W. Silverman and C. M. Theobald. Of course, any errors, omissions or obscurities which remain are entirely my responsibility. The author will be glad to hear from any reader who wishes to make constructive comments.

Finally, it is a pleasure to thank Jean Honebon, Doreen Faulds, Elizabeth Aplin and Sue Collins for typing the four editions of this book in such an efficient way.

Christopher Chatfield

School of Mathematical Sciences
University of Bath
Bath, Avon, BA2 7AY, UK

December 1988

Abbreviations and notation

AR	Autoregressive
MA	Moving average
ARMA	Mixed autoregressive and moving average
ARIMA	Autoregressive integrated moving average
ac.f.	Autocorrelation function
acv.f.	Autocovariance function
FFT	Fast Fourier transform
$N(\mu, \sigma^2)$	A normal distribution, mean μ, variance σ^2
χ_v^2	A chi-squared random variable with v degrees of freedom
∇	The difference operator such that $\nabla X_t = X_t - X_{t-1}$
B	The backward shift operator such that $BX_t = X_{t-1}$
E	Expected value or expectation
$\hat{X}(N, k)$	The k-step-ahead forecast of X_{N+k} made at time N

1

Introduction

A time series is a collection of observations made sequentially in time. Examples occur in a variety of fields, ranging from economics to engineering, and methods of analysing time series constitute an important area of statistics.

1.1 SOME REPRESENTATIVE TIME SERIES

We begin with some examples of the sort of time series which arise in practice.

(a) *Economic time series*

Many time series arise in economics. Examples include share prices on successive days, export totals in successive months, average incomes in successive months, company profits in successive years, and so on.

Figure 1.1 shows part of the classic Beveridge wheat price index series which consists of the average wheat price in nearly 50 places in various countries measured in successive years from 1500 to 1869. This series is of particular interest to economic historians. The complete series is tabulated by Anderson (1971).

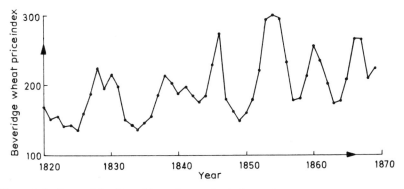

Figure 1.1 Part of the Beveridge wheat price index series.

(b) *Physical time series*

Many types of time series occur in the physical sciences, particularly in meteorology, marine science and geophysics. Examples are rainfall on successive days, and air temperature measured in successive hours, days or months. Figure 1.2 shows the air temperature at Recife, in Brazil, averaged over successive months. These data are tabulated and analysed in Example D.1.

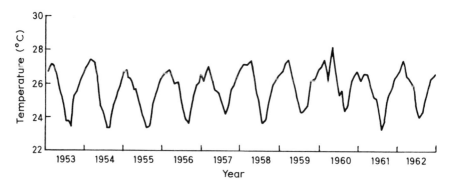

Figure 1.2   Average air temperature at Recife, Brazil, in successive months.

Some mechanical recorders take measurements continuously and produce a continuous trace rather than observations at discrete intervals of time. For example in some laboratories it is important to keep temperature and humidity as constant as possible and so devices are installed to measure these variables continuously. Some examples of continuous traces are given in Figure 7.4. In order to analyse such series, it may be helpful to sample or digitize them at equal intervals of time.

(c) *Marketing time series*

The analysis of sales figures in successive weeks or months is an important problem in commerce. Figure 1.3, taken from Chatfield and Prothero (1973), shows the sales of an engineering product by a certain company in successive months over a seven-year period. Marketing data have much in common with economic data. It is often important to forecast future sales so as to plan production. It may also be of interest to examine the relationship between sales and other time series such as advertising expenditure.

(d) *Demographic time series*

Time series occur in the study of population. An example is the population of England and Wales measured annually. Demographers want to predict

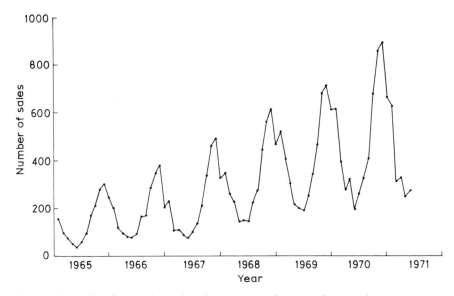

Figure 1.3 Sales of a certain engineering company in successive months.

changes in population for as long as ten or twenty years into the future (e.g. Brass, 1974).

(e) *Process control*

In process control, the problem is to detect changes in the performance of a manufacturing process by measuring a variable which shows the quality of the process. These measurements can be plotted against time as in Figure 1.4. When the measurements stray too far from some target value, appropriate corrective action should be taken to control the process. Special techniques

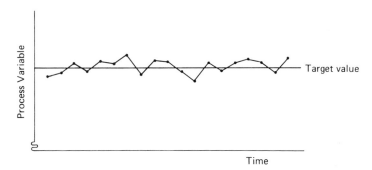

Figure 1.4 A process control chart.

have been developed for this type of time-series problem, and the reader is referred to a book on statistical quality control (e.g. Wetherill, 1977).

(f) *Binary processes*

A special type of time series arises when observations can take one of only two values, usually denoted by 0 and 1 (see Figure 1.5). Time series of this type, called binary processes, occur particularly in communication theory. For example the position of a switch, either 'on' or 'off', could be recorded as one or zero respectively.

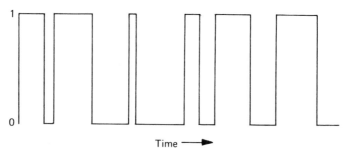

Figure 1.5 A realization of a binary process.

(g) *Point processes*

A different type of time series occurs when we consider a series of events occurring 'randomly' in time. For example we could record the dates of major railway disasters. A series of events of this type is often called a **point process** (see Figure 1.6). For observations of this type, we are interested in the distribution of the number of events occurring in a given time period and also in the distribution of time intervals between events. Methods of analysing data of this type will not be specifically discussed in this book (see for example Cox and Lewis, 1966; Cox and Isham, 1980).

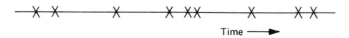

Figure 1.6 A realization of a point process (× denotes an event).

1.2 TERMINOLOGY

A time series is said to be **continuous** when observations are made continuously in time as in Figures 1.5 and 7.4. The term 'continuous' is used for series of this type even when the measured variable can only take a discrete set of values, as in Figure 1.5. A time series is said to be **discrete** when observations are taken only at specific times, usually equally spaced. The term 'discrete' is used for series of this type even when the measured variable is a continuous variable.

In this book we are mainly concerned with discrete time series, where the observations are taken at equal intervals. We also consider continuous time series more briefly, and in Section 11.8 we give some references regarding the analysis of discrete time series taken at unequal intervals of time.

Discrete time series can arise in several ways. Given a continuous time series, we could read off (or digitize) the values at equal intervals of time to give a discrete series called a **sampled** series. Another type of discrete series occurs when a variable does not have an instantaneous value but we can **aggregate** (or accumulate) the values over equal intervals of time. Examples of this type are exports measured monthly and rainfall measured daily. Finally, some time series are inherently discrete, an example being the dividend paid by a company to shareholders in successive years.

Much statistical theory is concerned with random samples of independent observations. The special feature of time-series analysis is the fact that successive observations are usually **not** independent and that the analysis must take into account the time **order** of the observations. When successive observations are dependent, future values may be predicted from past observations. If a time series can be predicted exactly, it is said to be **deterministic**. But most time series are **stochastic** in that the future is only partly determined by past values, so that exact predictions are impossible and must be replaced by the idea that future values have a probability distribution which is conditioned by a knowledge of past values.

1.3 OBJECTIVES OF TIME-SERIES ANALYSIS

There are several possible objectives in analysing a time series. These objectives may be classified as description, explanation, prediction and control, and will be considered in turn.

(a) *Description*

When presented with a time series, the first step in the analysis is usually to plot the data and to obtain simple descriptive measures of the main properties of the series as described in Chapter 2. For example, looking at Figure 1.3 it can be seen that there is a regular seasonal effect, with sales 'high' in winter and 'low' in summer. It also looks as though annual sales are increasing (i.e. show an upward trend). For some series, the variation is dominated by such 'obvious' features, and a fairly simple model, which only attempts to describe trend and seasonal variation, may be perfectly adequate to describe the variation in the time series. For other series, more sophisticated techniques will be required to provide an adequate analysis. Then a more complex model will be constructed, such as the various types of stochastic process described in Chapter 3.

This book devotes a greater amount of space to the more advanced techniques, but this does not mean that elementary descriptive techniques are unimportant. Anyone who tries to analyse a time series without plotting it first is asking for trouble. Not only will a graph show up trend and seasonal variation, but it also enables one to look for 'wild' observations or **outliers** which do not appear to be consistent with the rest of the data. The treatment of outliers is a complex subject in which common sense is as important as theory. The 'outlier' may be a perfectly valid but extreme observation which may for example indicate that the data are not normally distributed. Alternatively, the outlier may be a freak observation arising, for example, when a recording device goes wrong or when a strike severely affects sales. In the latter case, the outlier needs to be adjusted in some way before further analysis of the data. **Robust** methods (e.g. Martin, 1983) are designed to be insensitive to outliers.

Another feature to look for in the graph of the time series is the possible presence of turning points, where, for example, an upward trend has suddenly changed to a downward trend. If there is a turning point, different models may have to be fitted to the two parts of the series.

(b) *Explanation*

When observations are taken on two or more variables, it may be possible to use the variation in one time series to explain the variation in another series. This may lead to a deeper understanding of the mechanism which generated a given time series.

Multiple regression models may be helpful here. In Chapter 9 we also consider the analysis of what are called **linear systems**. A linear system converts an input series to an output series by a linear operation. Given observations on the input and output to a linear system (see Figure 1.7), one wants to assess the properties of the linear system. For example it is of interest to see how sea level is affected by temperature and pressure, and to see how sales are affected by price and economic conditions.

(c) *Prediction*

Gven an observed time series, one may want to predict the future values of the series. This is an important task in sales forecasting, and in the analysis of economic and industrial time series. Many writers, including myself, use the

Figure 1.7   Schematic representation of a linear system.

terms 'prediction' and 'forecasting' interchangeably, but some authors do not. For example Brown (1963) uses 'prediction' to describe subjective methods and 'forecasting' to describe objective methods, whereas Brass (1974) uses 'forecast' to mean any kind of looking into the future, and 'prediction' to denote a systematic procedure for doing so.

Prediction is closely related to **control** problems in many situations. For example if one can predict that a manufacturing process is going to move off target, then appropriate corrective action can be taken.

(d) *Control*

When a time series is generated which measures the 'quality' of a manufacturing process, the aim of the analysis may be to control the process. Control procedures are of several different kinds. In statistical quality control, the observations are plotted on control charts and the controller takes action as a result of studying the charts. A more sophisticated control strategy has been described by Box and Jenkins (1970). A stochastic model is fitted to the series, future values of the series are predicted, and then the input process variables are adjusted so as to keep the process on target. Many other contributions to control theory have been made by control engineers and mathematicians rather than statisticians. This topic is rather outside the scope of this book but is briefly introduced in Section 11.1.

1.4   APPROACHES TO TIME-SERIES ANALYSIS

This book will describe various approaches to time series analysis. In Chapter 2 we will describe simple descriptive techniques, which consist of plotting the data and looking for trends, seasonal fluctuations, and so on. Chapter 3 introduces a variety of probability models for time series, while Chapter 4 discusses ways of fitting these models to time series. The major diagnostic tool which is used in Chapter 4 is a function called the **autocorrelation** function which helps to describe the evolution of a process through time. Inference based on this function is often called an analysis in the **time domain**.

Chapter 5 discusses a variety of forecasting procedures. This chapter is not a prerequisite for the rest of the book and the reader may, if he wishes, proceed from Chapter 4 to Chapter 6.

Chapter 6 introduces a function called the **spectral density** function which describes how the variation in a time series may be accounted for by cyclic components at different frequencies. Chapter 7 shows how to estimate this function, a procedure which is called **spectral analysis**. Inference based on the spectral density function is often called an analysis in the **frequency domain**.

Chapter 8 discusses the analysis of two time series, while Chapter 9 extends this work by considering linear systems in which one series is regarded as the input, while the other series is regarded as the output.

Chapter 10 introduces a class of models, called state-space models, which are of much current interest.

Finally, Chapter 11 gives a brief introduction and references to a number of other topics, including some recent research developments.

1.5   REVIEW OF BOOKS ON TIME SERIES

This section gives a brief review of other books on time series. The literature has expanded considerably in recent years and a selective approach is necessary.

Introductory texts include Harvey (1981a), Gottman (1981), which is aimed primarily at social scientists, and Vandaele (1983), which concentrates on the Box-Jenkins modelling procedure (see Chapter 5).

A thorough, more mathematical, text for the univariate case is provided by Anderson (1971), while the comprehensive two-volume treatise by Priestley (1981) is particularly strong on spectral analysis, multivariate time series and non-linear models, and is a valuable reference source. The fourth edition of Kendall, Stuart and Ord (1983) is also a valuable reference source, but note that earlier editions are now rather dated.

Books on spectral analysis include Koopmans (1974), a clear introduction by Bloomfield (1976), and a text written primarily for engineers (Bendat and Piersol, 1986). The book by Jenkins and Watts (1968) is still a valuable reference source on spectral analysis, although it has been somewhat overtaken in parts by the development of the fast Fourier transform algorithm.

There are a number of books with a more theoretical flavour, including Quenouille (1957), Hannan (1970), Brillinger (1975) and Fuller (1976), and these are suitable for the more advanced reader.

The famous book by Box and Jenkins (1970) describes an approach to time-series analysis, forecasting and control which is based on a particular class of linear stochastic processes. While very important, this book is not entirely suitable for the beginner and an introduction to the Box-Jenkins approach is given in Chapters 3, 4, 5 and 9. Other useful books on forecasting include Gilchrist (1976), Montgomery and Johnson (1976), Abraham and Ledolter (1983) and Granger and Newbold (1986). Books on control theory are reviewed in Section 11.1.

2

Simple descriptive techniques

Statistical techniques for analysing time series range from relatively straight-forward **descriptive** methods to sophisticated **inferential** techniques. This chapter introduces the former, which will often clarify the main properties of a given series and should generally be tried anyway before attempting more complicated procedures.

2.1 TYPES OF VARIATION

Traditional methods of time-series analysis are mainly concerned with decomposing the variation in a series into trend, seasonal variation, other cyclic changes, and the remaining 'irregular' fluctuations. This approach is not always the best but is particularly valuable when the variation in dominated by trend and/or seasonality. However, it is worth noting that the decomposition is generally not unique unless certain assumptions are made. Thus some sort of modelling, either explicit or implicit, may be involved in these descriptive techniques, and this demonstrates the blurred borderline between descriptive and inferential techniques.

The different sources of variation will now be described in more detail.

(a) *Seasonal effect*

Many time series, such as sales figures and temperature readings, exhibit variation which is annual in period. For example, unemployment is typically 'high' in winter but lower in summer. This yearly variation is easy to understand, and we shall see that it can be measured explicitly and/or removed from the data to give deseasonalized data.

(b) *Other cyclic changes*

Apart from seasonal effects, some time series exhibit variation at a fixed period due to some other physical cause. An example is daily variation in temperature. In addition some time series exhibit oscillations which do not have a fixed period but which are predictable to some extent. For example economic data are sometimes thought to be affected by business cycles with a period varying between about 5 and 7 years, although the existence of such business cycles is the subject of some controversy.

(c) *Trend*

This may be loosely defined as 'long-term change in the mean level'. A difficulty with this definition is deciding what is meant by 'long term'. For example, climatic variables sometimes exhibit cyclic variation over a very long time period such as 50 years. If one just had 20 years' data, this long-term oscillation would appear to be a trend, but if several hundred years' data were available, the long-term oscillation would be visible. Nevertheless in the short term it may still be more meaningful to think of such a long-term oscillation as a trend. Thus in speaking of a 'trend', we must take into account the number of observations available and make a subjective assessment of what is 'long term'. Granger (1966) defines 'trend in mean' as comprising all cyclic components whose wavelength exceeds the length of the observed time series.

(d) *Other irregular fluctuations*

After trend and cyclic variations have been removed from a set of data, we are left with a series of residuals, which may or may not be 'random'. We shall examine various techniques for analysing series of this type to see if some of the apparently irregular variation may be explained in terms of probability models, such as **moving average** or **autoregressive** models which will be introduced in Chapter 3. Alternatively we can see if any cyclic variation is still left in the residuals.

2.2 STATIONARY TIME SERIES

A mathematical definition of a stationary time series will be given later on. However, it is now convenient to introduce the idea of stationarity from an intuitive point of view. Broadly speaking a time series is said to be **stationary** if there is no systematic change in mean (no trend), if there is no systematic change in variance, and if strictly periodic variations have been removed.

Most of the probability theory of time series is concerned with stationary time series, and for this reason time-series analysis often requires one to turn a non-stationary series into a stationary one so as to use this theory. For example it may be of interest to remove the trend and seasonal variation from a set of data and then try to model the variation in the residuals by means of a stationary stochastic process. However, it is also worth stressing that the non-stationary components, such as the trend, may sometimes be of more interest than the stationary residuals.

2.3 THE TIME PLOT

After getting background information and carefully defining objectives, the first, and most important, step in any time-series analysis is to plot the observations against time. This graph should show up important features of the series such as trend, seasonality, outliers and discontinuities. The plot is vital, both to describe the data and to help in formulating a sensible model, and a variety of examples are given throughout this book.

Plotting a time series is not as easy as it sounds. The choice of scales, the size of the intercept, and the way that the points are plotted (e.g. as a continuous line or as separate dots) may substantially affect the way the plot 'looks', and so the analyst must exercise care and judgement. In addition, the usual rules for drawing 'good' graphs should be followed: a clear title must be given, units of measurement should be stated, and axes should be clearly labelled.

Nowadays, graphs are often produced by computers. Some are well done but other packages may produce rather poor graphs and the reader must be prepared to modify them if necessary. For example, one package plotted the data in Figure 5.1(a) with a vertical scale labelled unhelpfully from '.4000E + 03' to '.2240E + 04'! The hideous E notation was naturally changed before publication.

2.4 TRANSFORMATIONS

Plotting the data may suggest that it is sensible to consider transforming them, for example by taking logarithms or square roots. The three main reasons for making a transformation are as follows.

(a) *To stabilize the variance*

If there is trend in the series and the variance appears to increase with the mean then it may be advisable to transform the data. In particular if the standard deviation is directly proportional to the mean, a logarithmic transformation is indicated.

(b) *To make the seasonal effect additive*

If there is a trend in the series and the size of the seasonal effect appears to increase with the mean then it may be advisable to transform the data so as to make the seasonal effect constant from year to year. The seasonal effect is then said to be additive. In particular if the size of the seasonal effect is directly proportional to the mean, then the seasonal effect is said to be multiplicative and a logarithmic transformation is appropriate to make the effect additive. However, this transformation will only stabilize the variance if the error term is **also** thought to be multiplicative (see Section 2.6), a point which is sometimes overlooked.

(c) *To make the data normally distributed*

Model building and forecasting are usually carried out on the assumption that the data are normally distrbuted. In practice this is not necessarily the case; there may for example be evidence of skewness in that there tend to be 'spikes' in the time plot which are all in the same direction (up or down). This effect can be difficult to eliminate and it may be necessary to assume a different 'error' distribution.

The logarithmic and square-root transformations are special cases of the class of transformations called the Box-Cox transformation. Given an observed time series $\{x_t\}$ and a transformation parameter λ, the transformed series is given by

$$y_t = \begin{cases} (x_t^\lambda - 1)/\lambda & \lambda \neq 0 \\ \log x_t & \lambda = 0 \end{cases}$$

This is effectively just a power transformation when $\lambda \neq 0$, as the constants are introduced to make y_t a continuous function of λ at the value $\lambda = 0$. The 'best' value of λ can be guesstimated, or alternatively estimated by a proper inferential procedure, such as maximum likelihood.

It is instructive to note that Nelson and Granger (1979) found little improvement in forecast performance when a general Box-Cox transformation was tried on a number of series. There are problems in practice with transformations in that a transformation which, say, makes the seasonal effect additive may fail to stabilize the variance and it may be impossible to achieve all requirements at the same time. In any case a model constructed for the transformed data may be less than helpful. For example, forecasts produced by the transformed model may have to be 'transformed back' in order to be of use and this can introduce biasing effects. My personal preference nowadays is to avoid transformations wherever possible except where the transformed variable has a direct physical interpretation. For example, when percentage increases are of interest, then taking logarithms makes sense (see Example D.3). Further general remarks on transformations are given by Granger and Newbold (1986, Section 10.5).

2.5   ANALYSING SERIES WHICH CONTAIN A TREND

In Section 2.1 we loosely defined trend as a 'long-term change in the mean level'. It is much more difficult to give a precise definition and different authors may use the term in different ways. The simplest trend is the familiar 'linear trend + noise', for which the observation at time t is a random variable X_t given by

$$X_t = \alpha + \beta t + \varepsilon_t \tag{2.1}$$

where α, β are constants and ε_t denotes a random error term with zero mean.

The mean level at time t is given by $m_t = (\alpha + \beta t)$; this is sometimes called 'the trend term'. Other writers prefer to describe the slope β as the trend; the trend is then the **change** in the mean level per unit time. It is usually clear from the context which meaning is intended.

The trend in equation (2.1) is a deterministic function of time and is sometimes called a **global** linear trend. This is generally unrealistic, and there is now more emphasis on **local** linear trends where the parameters α and β in equation (2.1) are allowed to evolve through time. Alternatively, the trend may be of non-linear form such as quadratic growth. Exponential growth can be particularly difficult to handle, even if logarithms are taken to transform the trend to a linear form.

The analysis of a time series which exhibits trend depends on whether one wants to (a) measure the trend and/or (b) remove the trend in order to analyse local fluctuations. It also depends on whether the data exhibit seasonality (see Section 2.6). With seasonal data, it is a good idea to start by calculating successive yearly averages as these will provide a simple description of the underlying trend. An approach of this type is sometimes perfectly adequate, particularly if the trend is fairly small, but sometimes a more sophisticated approach is desired and then the following techniques can be considered.

2.5.1 Curve fitting

A traditional method of dealing with non-seasonal data which contain a trend, particularly yearly data, is to fit a simple function such as a polynomial curve (linear, quadratic etc.), a Gompertz curve or a logistic curve (e.g. see Harrison and Pearce, 1972; Button, 1974; Levenbach and Reuter, 1976). The Gompertz curve is given by

$$\log x_t = a - br^t$$

where a, b, r are parameters with $0 < r < 1$, while the logistic curve is given by

$$x_t = a/(1 + b\,e^{-ct})$$

Both these curves are S-shaped and approach an asymptotic value as $t \to \infty$. Fitting the curves to data may lead to non-linear simultaneous equations.

For all curves of this type, the fitted function provides a measure of the trend, and the residuals provide an estimate of local fluctuations, where the residuals are the differences between the observations and the corresponding values of the fitted curve.

2.5.2 Filtering

A second procedure for dealing with a trend is to use a **linear filter** which converts one time series, $\{x_t\}$, into another, $\{y_t\}$, by the linear operation

$$y_t = \sum_{r=-q}^{+s} a_r x_{t+r}$$

where $\{a_r\}$ is a set of weights. In order to smooth out local fluctuations and estimate the local mean, we should clearly choose the weights so that $\Sigma a_r = 1$, and then the operation is often referred to as a **moving average**. Moving averages are discussed in detail by Kendall, Stuart and Ord (1983, Chapter 46), and we will only provide a brief introduction. Moving averages are often symmetric with $s = q$ and $a_j = a_{-j}$. The simplest example of a symmetric smoothing filter is the simple moving average, for which $a_r = 1/(2q+1)$ for $r = -q, \ldots, +q$, and the smoothed value of x_t is given by

$$\text{Sm}(x_t) = \frac{1}{2q+1} \sum_{r=-q}^{+q} x_{t+r}$$

The simple moving average is not generally recommended by itself for measuring trend, although it can be useful for removing seasonal variation. Another example is provided by the case where the $\{a_r\}$ are successive terms in the expansion of $(\frac{1}{2}+\frac{1}{2})^{2q}$. Thus when $q=1$, the weights are $a_{-1} = a_{+1} = \frac{1}{4}$, $a_0 = \frac{1}{2}$. As q gets large, the weights approximate to a normal curve.

A third example is Spencer's 15-point moving average, which is used for smoothing mortality statistics to get life tables (Tetley, 1946). This covers 15 consecutive points with $q=7$, and the symmetric weights are

$$\frac{1}{320} [-3, -6, -5, 3, 21, 46, 67, 74, \ldots]$$

A fourth example, called the Henderson moving average, is described by Kenny and Durbin (1982) and is finding increased use. This average aims to follow a cubic polynomial trend without distortion, and the choice of q depends on the degree of irregularity. The symmetric 9-term moving average, for example, is given by

$$[-0.041, -0.010, 0.119, 0.267, 0.330, \ldots]$$

The general idea is to fit a polynomial curve, not to the whole series, but to different parts. For example a polynomial fitted to the first $(2q+1)$ data points can be used to determine the interpolated value at the middle of the range where $t = (q+1)$, and the procedure can then be repeated using the data from $t=2$ to $t=(2q+2)$, and so on. A related idea is to use the class of piecewise polynomials called splines (e.g. Wegman and Wright, 1983).

Whenever a symmetric filter is chosen, there is likely to be an **end-effects** problem (e.g. Kendall, Stuart and Ord, 1983, Section 46.11), since $\text{Sm}(x_t)$ is calculated for $t = (q+1)$ to $t = N-q$. In some situations this is not important, but in other situations it is particularly important to get smoothed values up to

$t = N$. The analyst can project the smoothed values by eye, or by some further smoothing procedure, or, alternatively, use an asymmetric filter which only involves present and past values of x_t. For example, the popular technique known as exponential smoothing (see Section 5.2.2) effectively assumes that

$$\text{Sm}(x_t) = \sum_{j=0}^{\infty} \alpha(1-\alpha)^j x_{t-j}$$

where α is a constant such that $0 < \alpha < 1$. Here we note that the weights $a_j = \alpha(1-\alpha)^j$ decrease geometrically with j.

Having estimated the trend, we can look at the local fluctuations by examining

$$\text{Res}(x_t) = \text{residual from smoothed value}$$

$$= x_t - \text{Sm}(x_t)$$

$$= \sum_{r=-q}^{+s} b_r x_{t+r}$$

This is also a linear filter, and if $\Sigma a_r = 1$, then $\Sigma b_r = 0$, $b_0 = 1 - a_0$, and $b_r = -a_r$ for $r \neq 0$.

How do we choose the appropriate filter? The answer to this question really requires considerable experience plus a knowledge of the frequency aspects of time-series analysis which will be discussed in later chapters. As the name implies, filters are usually designed to produce an output with emphasis on variation at particular frequencies. For example, to get smoothed values we want to remove the local fluctuations which constitute what is called the high-frequency variation. In other words we want what is called a **low-pass** filter. To get $\text{Res}(x_t)$, we want to remove the long-term fluctuations or the low-frequency variation. In other words we want what is called a **high-pass** filter. The Slutsky (or Slutsky-Yule) effect is related to this problem. Slutsky showed that by operating on a completely random series with both averaging and differencing procedures one could induce sinusoidal variation in the data, and he went on to suggest that apparently periodic behaviour in some economic time series might be accounted for by the smoothing procedures used to form the data. We will return to this question later.

Filters in series

Very often a smoothing procedure is carried out in two or more stages – so that one has in effect several linear filters in series. For example two filters in series may be represented as follows:

Filter I with weights $\{a_{j1}\}$ acts on $\{x_t\}$ to produce $\{y_t\}$. Filter II with weights $\{a_{j2}\}$ acts on $\{y_t\}$ to produce $\{z_t\}$. Now

$$z_t = \sum_j a_{j2} y_{t+j}$$

$$= \sum_j a_{j2} \sum_r a_{r1} x_{t+j+r}$$

$$= \sum_j c_j x_{t+j}$$

where

$$c_j = \sum_r a_{r1} a_{(j-r)2}$$

are the weights for the overall filter. The weights $\{c_j\}$ are obtained by a procedure called **convolution**, and we write

$$\{c_j\} = \{a_{r1}\} * \{a_{r2}\}$$

where $*$ represents convolution. For example, the filter $(\frac{1}{4}, \frac{1}{2}, \frac{1}{4})$ may be written as

$$(\tfrac{1}{4}, \tfrac{1}{2}, \tfrac{1}{4}) = (\tfrac{1}{2}, \tfrac{1}{2}) * (\tfrac{1}{2}, \tfrac{1}{2})$$

Given a series $x_1, \ldots, x_N$, this smoothing procedure is best done in three stages by adding successive pairs of observations twice and then dividing by 4, as follows:

Observations	Stage I	Stage II	Stage III
x_1			
	$x_1 + x_2$		
x_2		$x_1 + 2x_2 + x_3$	$(x_1 + 2x_2 + x_3)/4$
	$x_2 + x_3$		.
x_3		$x_2 + 2x_3 + x_4$	.
	$x_3 + x_4$		.
x_4		$x_3 + 2x_4 + x_5$	.
	$x_4 + x_5$	.	.
x_5		.	.
	$x_5 + x_6$	.	.
x_6	.	.	.
.	.	.	.
.	.	.	.

The Spencer 15-point moving average is actually a convolution of four filters, namely

$$(\tfrac{1}{4}, \tfrac{1}{4}, \tfrac{1}{4}, \tfrac{1}{4}) * (\tfrac{1}{4}, \tfrac{1}{4}, \tfrac{1}{4}, \tfrac{1}{4}) * (\tfrac{1}{5}, \tfrac{1}{5}, \tfrac{1}{5}, \tfrac{1}{5}, \tfrac{1}{5}) * (-\tfrac{3}{4}, \tfrac{3}{4}, 1, \tfrac{3}{4}, -\tfrac{3}{4})$$

2.5.3 Differencing

A special type of filtering, which is particularly useful for removing a trend, is simply to difference a given time series until it becomes stationary. This method is an integral part of the procedures advocated by Box and Jenkins (1970). For non-seasonal data, first-order differencing is usually sufficient to attain apparent stationarity, so that the new series $\{y_1, \ldots, y_{N-1}\}$ is formed from the original series $\{x_1, \ldots, x_N\}$ by

$$y_t = x_{t+1} - x_t = \nabla x_{t+1}$$

First-order differencing is widely used. Occasionally second-order differencing is required using the operator ∇^2, where

$$\nabla^2 x_{t+2} = \nabla x_{t+2} - \nabla x_{t+1} = x_{t+2} - 2x_{t+1} + x_t$$

2.6 ANALYSING SERIES WHICH CONTAIN SEASONAL VARIATION

In Section 2.1 we introduced seasonal variation which is generally annual in period, while Section 2.4 distinguished between additive seasonality, which is constant from year to year, and multiplicative seasonality. Three seasonal models in common use are

A $X_t = m_t + S_t + \varepsilon_t$

B $X_t = m_t S_t + \varepsilon_t$

C $X_t = m_t S_t \varepsilon_t$

where m_t is the deseasonalized mean level at time t, S_t is the seasonal effect at time t, and ε_t is the random error.

Model A describes the additive case, while models B and C both involve multiplicative seasonality. In model C the error term is also multiplicative, and a logarithmic transformation will turn this into a (linear) additive model which may be easier to handle. The time plot should be examined to see which model is likely to give the better description. The seasonal indices $\{S_t\}$ are usually assumed to change slowly through time so that $S_t \simeq S_{t-s}$, where s is the number of observations per year. The indices are usually normalized so that they sum to zero in the additive case, or average to one in the multiplicative case. Difficulties arise in practice if the seasonal and/or error terms are not exactly multiplicative or additive. For example the seasonal effect may increase with the mean level but not at such a fast rate so that it is somewhere 'in between' being multiplicative or additive. A mixed additive-multiplicative seasonal model is described by Durbin and Murphy (1975).

The analysis of time series which exhibit seasonal variation depends on whether one wants to (a) measure the seasonal effect and/or (b) eliminate seasonality. For series showing little trend, it is usually adequate to estimate the seasonal effect for a particular period (e.g. January) by finding the average of each January observation minus the corresponding yearly average in the additive case, or the January observation divided by the yearly average in the multiplicative case.

For series which do contain a substantial trend, a more sophisticated approach may be required. With monthly data, the commonest way of eliminating the seasonal effect is to calculate

$$\text{Sm}(x_t) = \frac{\frac{1}{2}x_{t-6} + x_{t-5} + x_{t-4} + \cdots + x_{t+5} + \frac{1}{2}x_{t+6}}{12}$$

Note that the sum of the coefficients is 1. A simple moving average cannot be used as this would span 12 months and would not be centred on an integer value of t. A simple moving average over 13 months cannot be used, as this would give twice as much weight to the month appearing at both ends. For quarterly data, the seasonal effect can be eliminated by calculating

$$\text{Sm}(x_t) = \frac{\frac{1}{2}x_{t-2} + x_{t-1} + x_t + x_{t+1} + \frac{1}{2}x_{t+2}}{4}$$

For 4-weekly data, one **can** use a simple moving average over 13 successive observations. The seasonal effect can be estimated by calculating $x_t - \text{Sm}(x_t)$ or $x_t/\text{Sm}(x_t)$ depending on whether the seasonal effect is thought to be additive or multiplicative. A check should be made that the seasonals are reasonably stable, and then the average monthly (or quarterly etc.) effects can be calculated.

A seasonal effect can also be eliminated by differencing (see Sections 4.6, 5.2.4; Box and Jenkins, 1970). For example with monthly data one can employ the operator ∇_{12} where

$$\nabla_{12}x_t = x_t - x_{t-12}$$

Alternative methods of seasonal adjustment are reviewed by Pierce (1980), Cleveland (1983) and Newbold (1984). These include the widely used X-11 method which employs a series of linear filters (e.g. see Wallis, 1982; Kendall, Stuart and Ord, 1983, Section 46.41). The possible presence of calendar effects should also be considered (Cleveland and Devlin, 1982; Cleveland, 1983). For example if Easter falls in March one year, rather than April, then this may alter the seasonal effect on sales for both months.

2.7   AUTOCORRELATION

An important guide to the properties of a time series is provided by a series of quantities called sample autocorrelation coefficients, which measure the correlation between observations at different distances apart. These coefficients often provide insight into the probability model which generated the data. We assume that the reader is familiar with the ordinary correlation coefficient, namely that given N pairs of observations on two variables x and y, the correlation coefficient is given by

$$r = \frac{\Sigma(x_i - \bar{x})(y_i - \bar{y})}{\sqrt{[\Sigma(x_i - \bar{x})^2 \Sigma(y_i - \bar{y})^2]}} \tag{2.2}$$

A similar idea can applied to time series to see if successive observations are correlated.

Given N observations $x_1, \ldots, x_N$, on a discrete time series we can form $N-1$ pairs of observations, namely $(x_1, x_2), (x_2, x_3), \ldots, (x_{N-1}, x_N)$. Regarding the first observation in each pair as one variable, and the second observation as a second variable, the correlation coefficient between x_t and x_{t+1} is given by

$$r_1 = \frac{\sum_{t=1}^{N-1} (x_t - \bar{x}_{(1)})(x_{t+1} - \bar{x}_{(2)})}{\sqrt{\left[\sum_{t=1}^{N-1} (x_t - \bar{x}_{(1)})^2 \sum_{t=1}^{N-1} (x_{t+1} - \bar{x}_{(2)})^2\right]}} \tag{2.3}$$

by analogy with equation (2.2), where

$$\bar{x}_{(1)} = \sum_{t=1}^{N-1} x_t/(N-1)$$

is the mean of the first $N-1$ observations and

$$\bar{x}_{(2)} = \sum_{t=2}^{N} x_t/(N-1)$$

is the mean of the last $N-1$ observations. As the coefficient given by equation (2.3) measures correlation between successive observations it is called an **autocorrelation** coefficient or serial correlation coefficient.

Equation (2.3) is rather complicated, and so, as $\bar{x}_{(1)} \simeq \bar{x}_{(2)}$, it is usually approximated by

$$r_1 = \frac{\sum_{t=1}^{N-1} (x_t - \bar{x})(x_{t+1} - \bar{x})}{(N-1)\sum_{t=1}^{N} (x_t - \bar{x})^2/N} \tag{2.4}$$

where $\bar{x} = \sum_{t=1}^{N} x_t/N$ is the overall mean. Some authors also drop the factor $N/(N-1)$, which is close to one for large N, to give the even simpler formula

$$r_1 = \frac{\sum\limits_{t=1}^{N-1} (x_t - \bar{x})(x_{t+1} - \bar{x})}{\sum\limits_{t=1}^{N} (x_t - \bar{x})^2} \qquad (2.5)$$

and this is the form that will be used in this book.

In a similar way we can find the correlation between observations a distance k apart, which is given by

$$r_k = \frac{\sum\limits_{t=1}^{N-k} (x_t - \bar{x})(x_{t+k} - \bar{x})}{\sum\limits_{t=1}^{N} (x_t - \bar{x})^2} \qquad (2.6)$$

This is called the autocorrelation coefficient at lag k.

In practice the autocorrelation coefficients are usually calculated by computing the series of autocovariance coefficients, $\{c_k\}$, which we define by analogy with the usual covariance formula as

$$c_k = \frac{1}{N} \sum\limits_{t=1}^{N-k} (x_t - \bar{x})(x_{t+k} - \bar{x}) \qquad (2.7)$$

This is the autocovariance coefficient at lag k.

We then compute

$$r_k = c_k / c_0 \qquad (2.8)$$

for $k = 1, 2, \ldots, m$, where $m < N$. There is often little point in calculating r_k for values of k greater than about $N/4$.

Note that some authors prefer to use

$$c_k = \frac{1}{N-k} \sum\limits_{t=1}^{N-k} (x_t - \bar{x})(x_{t+k} - \bar{x})$$

rather than equation 2.7, but there is little difference for large N (see Section 4.1).

2.7.1 The correlogram

A useful aid in interpreting a set of autocorrelation coefficients is a graph called a correlogram in which r_k is plotted against the lag k. Examples are given in Figures 2.1–2.3. A visual inspection of the correlogram is often very helpful.

2.7.2 Interpreting the correlogram

Interpreting the meaning of a set of autocorrelation coefficients is not always easy. Here we offer some general advice.

(a) *A random series*

If a time series is completely random, then for large N, $r_k \simeq 0$ for all non-zero values of k. In fact we will see later that for a random time series r_k is approximately $N(0, 1/N)$, so that, if a time series is random, 19 out of 20 of the values of r_k can be expected to lie between $\pm 2/\sqrt{N}$. However if one plots say the first 20 values of r_k then one can expect to find one 'significant' value on average even when the time series really is random. This spotlights one of the difficulties in interpreting the correlogram, in that a large number of coefficients is quite likely to contain one (or more) 'unusual' results, even when no real effects are present. (See also Section 4.1.)

(b) *Short-term correlation*

Stationary series often exhibit short-term correlation characterized by a fairly large value of r_1 followed by a few further coefficients which, while greater than zero, tend to get successively smaller. Values of r_k for longer lags tend to be approximately zero. An example of such a correlogram is shown in Figure 2.1. A time series which gives rise to such a correlogram is one for which an observation above the mean tends to be followed by one or more further observations above the mean, and similarly for observations below the mean.

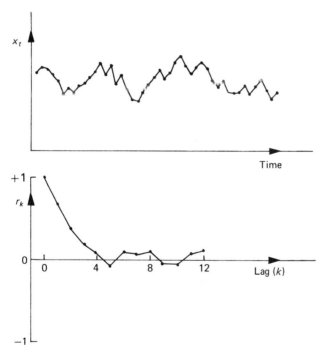

Figure 2.1   A time series showing short-term correlation together with its correlogram.

(c) *Alternating series*

If a time series has a tendency to alternate, with successive observations on different sides of the overall mean, then the correlogram also tends to alternate. The value of r_1 will be negative. However the value of r_2 will be positive, as observations at lag 2 will tend to be on the same side of the mean. A typical alternating time series together with its correlogram is shown in Figure 2.2.

(d) *Non-stationary series*

If a time series contains a trend, then the values of r_k will not come down to zero except for very large values of the lag. This is because an observation on one side of the overall mean tends to be followed by a large number of further observations on the same side of the mean because of the trend. A typical non-stationary time series together with its correlogram is shown in Figure 2.3. Little can be inferred from a correlogram of this type as the trend dominates all other features. In fact the sample autocorrelation function, $\{r_k\}$, is only meaningful for **stationary** time series (see Chapters 3 and 4) and so any trend should be removed before calculating $\{r_k\}$.

(e) *Seasonal fluctuations*

If a time series contains a seasonal fluctuation, then the correlogram will also exhibit an oscillation at the same frequency. For example with monthly observations, r_6 will be 'large' and negative, while r_{12} will be 'large' and

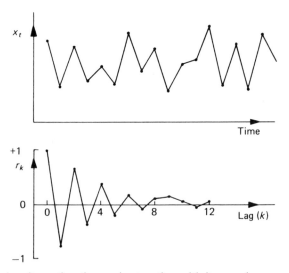

Figure 2.2   An alternating time series together with its correlogram.

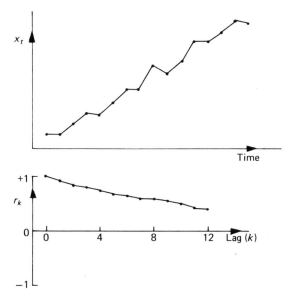

Figure 2.3   A non-stationary time series together with its correlogram.

positive. In particular if x_t follows a sinusoidal pattern, then so does r_k. For example, if

$$x_t = a \cos t\omega$$

where a is a constant and the frequency ω is such that $0 < \omega < \pi$, then it can be shown (see Exercise 2.3) that

$$r_k \simeq \cos k\omega \qquad \text{for large } N$$

Figure 2.4(a) shows the correlogram of the monthly air temperature data shown in Figure 1.2. The sinusoidal pattern of the correlogram is clearly evident, but for seasonal data of this type the correlogram provides little extra information as the seasonal pattern is clearly evident in the time plot of the data.

If the seasonal variation is removed from seasonal data, then the correlogram may provide useful information. The seasonal variation was removed from the air temperature data by the simple procedure of calculating the 12 monthly averages and subtracting the appropriate one from each individual observation. The correlogram of the resulting series (Figure 2.4(b)) shows that the first three coefficients are significantly different from zero. This indicates short-term correlation in that a month which is say colder than the average for that month will tend to be followed by one or two further months which are colder than average.

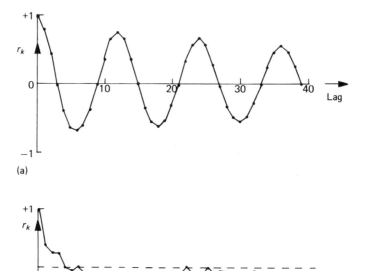

Figure 2.4   The correlogram of monthly observations on air temperature at Recife: (a) for the raw data; (b) for the seasonally adjusted data. The dotted lines in (b) are at $\pm 2/\sqrt{N}$. Values outside these lines are significantly different from zero.

(f) *Outliers*

If a time series contains one or more outliers, the correlogram may be seriously affected and it may be advisable to adjust outliers in some way before starting the formal analysis. For example, if there is one outlier in the time series and it is not adjusted, then the plot of x_t against x_{t+k} will contain **two** 'extreme' points which will tend to depress the sample correlation coefficients towards zero. If there are two outliers this effect is even more noticeable, except when the lag equals the distance between the outliers when a spuriously large correlation may occur.

(g) *General remarks*

Clearly considerable experience is required in interpreting autocorrelation coefficients. In addition we need to study the probability theory of stationary series and discuss the classes of model which may be appropriate. We must also discuss the sampling properties of r_k. These topics will be covered in the next two chapters and we shall then be in a better position to interpret the correlogram of a given time series.

2.8   OTHER TESTS OF RANDOMNESS

In most cases, a visual examination of the graph of a time series is enough to see that it is **not** random. However it is occasionally desirable to test a stationary time series for 'randomness'. In other words one wants to test if $x_1, \ldots, x_N$ could have arisen in that order by chance by taking a simple random sample size N from a population with unknown characteristics. A variety of tests exist for this purpose and they are described by Kendall, Stuart and Ord (1983, Section 45.15). For example one can examine the number of times there is a local maximum or minimum in the time series. A local maximum is defined to be any observation x_t such that $x_t > x_{t-1}$ and $x_t > x_{t+1}$. A converse definition applies to local minima. If the series really is random one can work out the expected number of turning points and compare it with the observed value. Tests of this type will not be described here, as I have always found it more convenient to simply examine the correlogram (and possibly the spectral density function) of a given time series to see if it is random.

Having fitted a model to a non-random series, one often wants to see if the residuals are random. Testing residuals for randomness is a somewhat different problem and will be discussed in Section 4.7.

EXERCISES

2.1   The following data show the sales of company X in successive 4-week periods over 1967 1970.

	I	II	III	IV	V	VI	VII	VIII	IX	X	XI	XII	XIII
1967	153	189	221	215	302	223	201	173	121	106	86	87	108
1968	133	177	241	228	283	255	238	164	128	108	87	74	95
1969	145	200	187	201	292	220	233	172	119	81	65	76	74
1970	111	170	243	178	248	202	163	139	120	96	95	53	94

(a) Plot the data.

(b) Assess the trend and seasonal effects.

2.2   Sixteen successive observations on a stationary time series are as follows:
1.6, 0.8, 1.2, 0.5, 0.9, 1.1, 1.1, 0.6, 1.5, 0.8, 0.9, 1.2, 0.5, 1.3, 0.8, 1.2

(a) Plot the observations.

(b) Looking at the graph, guess an approximate value for the autocorrelation coefficient at lag 1.

(c) Plot x_t against x_{t+1}, and again try to guess the value of r_1.

(d) Calculate r_1.

2.3   If $x_t = a \cos t\omega$ where a is a constant and ω is a constant in $(0, \pi)$, show that $r_k \to \cos k\omega$ as $N \to \infty$.

(Hint: You will need to use the trigonometrical results listed in Section 7.2. Using equation (7.2) it can be shown that $\bar{x} \to 0$ as $N \to \infty$, so that $r_k \to \Sigma \cos \omega t \cos \omega(t+k)/\Sigma \cos^2 \omega t$. Now use the result that $2 \cos A \cos B = \cos(A+B) + \cos(A-B)$ together with the result that $\Sigma \cos^2 \omega t = N/2$ for a suitably chosen N.)

2.4   The first ten sample autocorrelation coefficients of 400 'random' numbers are $r_1 = 0.02$, $r_2 = 0.05$, $r_3 = -0.09$, $r_4 = 0.08$, $r_5 = -0.02$, $r_6 = 0.00$, $r_7 = 0.12$, $r_8 = 0.06$, $r_9 = 0.02$, $r_{10} = -0.08$. Is there any evidence of non-randomness?

2.5   Given a seasonal series of monthly observations $\{X_t\}$, assume that the seasonal factors $\{S_t\}$ are constant so that $S_t = S_{t-12}$ for all t, and also that $\{\varepsilon_t\}$ is a stationary series of random deviations.

(a) With a global linear trend and additive seasonality, we have $X_t = a + bt + S_t + \varepsilon_t$. Show that the operator ∇_{12} acting on X_t reduces the series to stationarity.

(b) With a global linear trend and multiplicative seasonality, we have $X_t = (a + bt)S_t + \varepsilon_t$. Does the operator ∇_{12} reduce X_t to stationarity? If not, find a differencing operator which does.

(Note: As stationarity is not formally defined until Chapter 3, you should use heuristic arguments. A stationary process may involve a non-zero, but constant, mean value and any linear combination of the $\{\varepsilon_t\}$, but not terms such as S_t.)

3
Probability models for time series

3.1 STOCHASTIC PROCESSES

This chapter describes various probability models for time series, which are collectively called stochastic processes. Most physical processes in the real world involve a random element in their structure, and a **stochastic process** can be described as 'a statistical phenomenon that evolves in time according to probabilistic laws'. Well-known examples are the length of a queue, the size of a bacterial colony, and the air temperature on successive days at a particular site. The word 'stochastic', which is of Greek origin, is used to mean 'pertaining to chance', and many writers use 'random process' as a synonym for stochastic process.

Mathematically, a stochastic process may be defined as a collection of random variables which are ordered in time and defined at a set of time points which may be continuous or discrete. We will denote the random variable at time t by $X(t)$ if time is continuous (usually $-\infty < t < \infty$), and by X_t if time is discrete (usually $t = 0, \pm 1, \pm 2, \ldots$).

The theory of stochastic processes has been extensively developed and is discussed in many books including Papoulis (1984), written primarily for engineers, Parzen (1962), Cox and Miller (1968, especially Chapter 7), Yaglom (1962) and Grimmett and Stirzaker (1982). In this chapter we concentrate on those aspects particularly relevant to time-series analysis.

Most statistical problems are concerned with estimating the properties of a population from a sample. In time-series analysis there is a rather different situation in that, although it may be possible to vary the **length** of the observed time series – the sample – it is usually impossible to make more than one observation at any given time. Thus we only have a single outcome of the process and a single observation on the random variable at time t. Nevertheless we may regard the observed time series as just one example of the infinite set of time series which might have been observed. This infinite set of time series is sometimes called the **ensemble**. Every member of the ensemble is a possible **realization** of the stochastic process. The observed time series can be thought of as one particular realization, and will be denoted by $x(t)$ for

$(0 \leqslant t \leqslant T)$ if observations are continuous, and by x_t for $t = 1, \ldots, N$ if observations are discrete.

Because there is only a notional population, time-series analysis is essentially concerned with evaluating the properties of the probability model which generated the observed time series.

One way of describing a stochastic process is to specify the joint probability distribution of $X(t_1), \ldots, X(t_n)$ for any set of times $t_1, \ldots, t_n$ and any value of n. But this is rather complicated and is not usually attempted in practice. A simpler, more useful way of describing a stochastic process is to give the **moments** of the process, particularly the first and second moments, which are called the mean, variance and autocovariance functions. These will now be defined for continuous time, with similar definitions applying in discrete time.

Mean   The mean function $\mu(t)$ is defined by

$$\mu(t) = E[X(t)]$$

Variance   The variance function $\sigma^2(t)$ is defined by

$$\sigma^2(t) = \mathrm{Var}[X(t)]$$

Autocovariance   The variance function alone is not enough to specify the second moments of a sequence of random variables. In addition, we must define the autocovariance function $\gamma(t_1, t_2)$, which is the covariance of $X(t_1)$ with $X(t_2)$, namely

$$\gamma(t_1, t_2) = E\{[X(t_1) - \mu(t_1)][X(t_2) - \mu(t_2)]\}$$

(Readers who are unfamiliar with the term 'covariance' should read Appendix C. When applied to a sequence of random variables, it is called an autocovariance.) Note that the variance function is a special case of the autocovariance function when $t_1 = t_2$.

Higher moments of a stochastic process may be defined in an obvious way, but are rarely used in practice, since a knowledge of the two functions $\mu(t)$ and $\gamma(t_1, t_2)$ is usually adequate.

3.2   STATIONARY PROCESSES

An important class of stochastic processes are those which are stationary. A heuristic idea of stationarity was introduced in Section 2.2.

A time series is said to be **strictly stationary** if the joint distribution of $X(t_1), \ldots, X(t_n)$ is the same as the joint distribution of $X(t_1 + \tau), \ldots, X(t_n + \tau)$ for all $t_1, \ldots, t_n, \tau$. In other words, shifting the time origin by an amount τ has no effect on the joint distributions, which must therefore depend only on the intervals between $t_1, t_2, \ldots, t_n$. The above definition holds for any value of n.

In particular, if $n=1$ it implies that the distribution of $X(t)$ must be the same for all t, so that

$$\mu(t)=\mu$$
$$\sigma^2(t)=\sigma^2$$

are both constants which do not depend on the value of t.

Furthermore, if $n=2$ the joint distribution of $X(t_1)$ and $X(t_2)$ depends only on (t_2-t_1), which is called the **lag**. Thus the autocovariance function $\gamma(t_1, t_2)$ also depends only on (t_2-t_1) and may be written as $\gamma(\tau)$, where

$$\gamma(\tau)=E\{[X(t)-\mu][X(t+\tau)-\mu]\}$$
$$=\mathrm{Cov}[X(t), X(t+\tau)]$$

is called the autocovariance coefficient at lag τ. In future, 'autocovariance function' will be abbreviated to acv.f.

The size of an autocovariance coefficient depends on the units in which $X(t)$ is measured. Thus, for interpretative purposes, it is useful to standardize the acv.f. to produce a function called the **autocorrelation** function, which is given by

$$\rho(\tau)=\gamma(\tau)/\gamma(0)$$

and which measures the correlation between $X(t)$ and $X(t+\tau)$. Its empirical counterpart was introduced in Section 2.7. In future, 'autocorrelation function' will be abbreviated to ac.f. Note that the argument τ of $\gamma(\tau)$ and $\rho(\tau)$ is discrete if the time series is discrete and continuous if the time units is continuous.

At first sight it may seem surprising to suggest that there are processes for which the distribution of $X(t)$ should be the same for all t. However, readers with some knowledge of stochastic processes will know that there are many processes $\{X(t)\}$ which have what is called an **equilibrium** distribution as $t\to\infty$, in which the probability distribution of $X(t)$ tends to a limit which does **not** depend on the initial conditions. Thus once such a process has been running for some time, the distribution of $X(t)$ will change very little. Indeed if the initial conditions are specified to be identical to the equilibrium distribution, the process is stationary in time and the equilibrium distribution is then the stationary distribution of the process. Of course the **conditional** distribution of $X(t_2)$ given that $X(t_1)$ has taken a particular value, say $x(t_1)$, may be quite different from the stationary distribution, but this is perfectly consistent with the process being stationary.

3.2.1 Second-order stationarity

In practice it is often useful to define stationarity in a less restricted way than that described above. A process is called second-order stationary (or weakly stationary) if its mean is constant and its acv.f. depends only on the lag, so that

$$E[X(t)] = \mu$$

and

$$\text{Cov}[X(t), X(t+\tau)] = \gamma(\tau)$$

No assumptions are made about higher moments than those of second order. By letting $\tau = 0$, we note that the above assumption about the acv.f. implies that the variance, as well as the mean, is constant. Also note that both the variance and the mean must be finite.

   This weaker definition of stationarity will generally be used from now on, as many of the properties of stationary processes depend only on the structure of the process as specified by its first and second moments. One important class of processes where this is particularly true is the class of **normal** processes where the joint distribution of $X(t_1), \ldots, X(t_n)$ is multivariate normal for all $t_1, \ldots, t_n$. The multivariate normal distribution is completely characterized by its first and second moments, and hence by $\mu(t)$ and $\gamma(t_1, t_2)$, and so it follows that second-order stationarity implies strict stationarity for normal processes. However, μ and $\gamma(\tau)$ may not adequately describe processes which are very 'non-normal'.

3.3   THE AUTOCORRELATION FUNCTION

We have already noted in Section 2.7 that the sample autocorrelation coefficients of an observed time series are an important set of statistics for describing the time series. Similarly the (theoretical) autocorrelation function (ac.f.) of a stationary stochastic process is an important tool for assessing its properties. This section investigates the general properties of the ac.f.

   Suppose a stationary stochastic process $X(t)$ has mean μ, variance σ^2, acv.f. $\gamma(\tau)$, and ac.f. $\rho(\tau)$. Then

$$\rho(\tau) = \gamma(\tau)/\gamma(0) = \gamma(\tau)/\sigma^2$$

Note that $\rho(0) = 1$.

Property 1

The ac.f. is an **even** function of the lag in that

$$\rho(\tau) = \rho(-\tau)$$

This property simply says that the correlation between $X(t)$ and $X(t+\tau)$ is the same as that between $X(t)$ and $X(t-\tau)$. The result is easily proved using $\gamma(\tau) = \rho(\tau)\sigma^2$ by

$$\gamma(\tau) = \text{Cov}[X(t), X(t+\tau)]$$

$$= \text{Cov}[X(t-\tau), X(t)] \qquad \text{since } X(t) \text{ stationary}$$

$$= \gamma(-\tau)$$

Property 2

$|\rho(\tau)| \leqslant 1$. This is the 'usual' property of a correlation. It is proved by noting that

$$\text{Var}[\lambda_1 X(t) + \lambda_2 X(t+\tau)] \geqslant 0$$

for any constants λ_1, λ_2, since a variance is always non-negative. This variance is equal to

$$\lambda_1^2 \text{Var}[X(t)] + \lambda_2^2 \text{Var}[X(t+\tau)] + 2\lambda_1\lambda_2 \text{Cov}[X(t), X(t+\tau)]$$

$$= (\lambda_1^2 + \lambda_2^2)\sigma^2 + 2\lambda_1\lambda_2\gamma(\tau)$$

When $\lambda_1 = \lambda_2 = 1$, we find

$$\gamma(\tau) \geqslant -\sigma^2$$

so that $\rho(\tau) \geqslant -1$. When $\lambda_1 = 1, \lambda_2 = -1$, we find

$$\sigma^2 \geqslant \gamma(\tau)$$

so that $\rho(\tau) \leqslant +1$.

Property 3

Lack of uniqueness. Although a given stochastic process has a unique covariance structure, the converse is not in general true. It is usually possible to find many normal and non-normal processes with the same ac.f. and this creates further difficulty in interpreting sample ac.f.s. Jenkins and Watts (1968, p. 170) give an example of two different stochastic processes which have the same ac.f. Even for stationary normal processes, which are completely determined by the mean, variance and ac.f., the invertibility condition introduced in Section 3.4.3 is required to ensure uniqueness.

3.4   SOME USEFUL STOCHASTIC PROCESSES

This section describes several different types of stochastic process which are sometimes useful in setting up a model for a time series.

3.4.1   A purely random process

A discrete-time process is called a purely random process if it consists of a sequence of random variables $\{Z_t\}$ which are mutually independent and identically distributed. From the definition it follows that the process has constant mean and variance and that

$$\gamma(k) = \text{Cov}(Z_t, Z_{t+k})$$
$$= 0 \qquad \text{for } k = \pm 1, 2, \ldots$$

As the mean and acv.f. do not depend on time, the process is second-order stationary. In fact it is clear that the process is also strictly stationary. The ac.f. is given by

$$\rho(k) = \begin{cases} 1 & k = 0 \\ 0 & k = \pm 1, \pm 2, \ldots \end{cases}$$

A purely random process is sometimes called **white noise**, particularly by engineers. Processes of this type are useful in many situations, particularly as building blocks for more complicated processes such as moving average processes (Section 3.4.3).

The possibility of defining a continuous-time purely random process is discussed in Section 3.4.8.

3.4.2   Random walk

Suppose that $\{Z_t\}$ is a discrete, purely random process with mean μ and variance σ_Z^2. A process $\{X_t\}$ is said to be a random walk if

$$X_t = X_{t-1} + Z_t \tag{3.1}$$

The process is customarily started at zero when $t = 0$, so that

$$X_1 = Z_1$$

and

$$X_t = \sum_{i=1}^{t} Z_i$$

Then we find that $E(X_t) = t\mu$ and that $\text{Var}(X_t) = t\sigma_Z^2$. As the mean and variance change with t, the process is non-stationary.

However, it is interesting to note that the first differences of a random walk, given by

$$\nabla X_t = X_t - X_{t-1} = Z_t$$

form a purely random process, which is therefore stationary.

The best-known examples of time series which behave like random walks are share prices on successive days. A model which often gives a good approximation to such data is

share price on day t = share price on day $(t-1)$ + random error

3.4.3   Moving average processes

Suppose that $\{Z_t\}$ is a purely random process with mean zero and variance σ_Z^2. Then a process $\{X_t\}$ is said to be a moving average process of order q (abbreviated to an MA(q) process) if

$$X_t = \beta_0 Z_t + \beta_1 Z_{t-1} + \cdots + \beta_q Z_{t-q} \qquad (3.2)$$

where $\{\beta_i\}$ are constants. The Zs are usually scaled so that $\beta_0 = 1$.
    We find immediately that

$$E(X_t) = 0$$

$$\mathrm{Var}(X_t) = \sigma_Z^2 \sum_{i=0}^{q} \beta_i^2$$

since the Zs are independent. We also have

$$
\begin{aligned}
\gamma(k) = {} & \mathrm{Cov}(X_t, X_{t+k}) \\
= {} & \mathrm{Cov}(\beta_0 Z_t + \cdots + \beta_q Z_{t-q}, \beta_0 Z_{t+k} + \cdots + \beta_q Z_{t+k-q}) \\
= {} &
\begin{cases}
0 & k > q \\
\sigma_Z^2 \displaystyle\sum_{i=0}^{q-k} \beta_i \beta_{i+k} & k = 0, 1, \ldots, q \\
\gamma(-k) & k < 0
\end{cases}
\end{aligned}
$$

since

$$\mathrm{Cov}(Z_s, Z_t) = \begin{cases} \sigma_Z^2 & s = t \\ 0 & s \neq t \end{cases}$$

As $\gamma(k)$ does not depend on t, and the mean is constant, the process is second-order stationary for all values of the $\{\beta_i\}$. Furthermore, if the Zs are normally distributed, then so are the Xs, and we have a strictly stationary normal process.
    The ac.f. of the MA(q) process is given by

$$
\rho(k) =
\begin{cases}
1 & k = 0 \\
\displaystyle\sum_{i=0}^{q-k} \beta_i \beta_{i+k} \bigg/ \sum_{i=0}^{q} \beta_i^2 & k = 1, \ldots, q \\
0 & k > q \\
\rho(-k) & k < 0
\end{cases}
$$

Note that the ac.f. 'cuts off' at lag q, which is a special feature of MA processes.

In particular the MA(1) process with $\beta_0 = 1$ has an ac.f. given by

$$
\rho(k) = \begin{cases} 1 & k=0 \\ \beta_1/(1+\beta_1^2) & k=\pm 1 \\ 0 & \text{otherwise} \end{cases}
$$

Although no restrictions on the $\{\beta_i\}$ are required for an MA process to be stationary, it is generally desirable to impose restrictions on the $\{\beta_i\}$ to ensure that the process satisfies a condition called **invertibility** (e.g. Box and Jenkins, 1970, p. 50). This condition may be explained in the following way. Consider the following first-order MA processes:

$$
\text{A} \qquad X_t = Z_t + \theta Z_{t-1}
$$

$$
\text{B} \qquad X_t = Z_t + \frac{1}{\theta} Z_{t-1}
$$

It can easily be shown that these two different processes have exactly the same ac.f. (Are you surprised? Then check $\rho(k)$ for models A and B.) Thus we cannot identify an MA process uniquely from a given ac.f. Now, if we express models A and B by putting Z_t in terms of $X_t, X_{t-1}, \ldots$, we find by successive substitution that

$$
\text{A} \qquad Z_t = X_t - \theta X_{t-1} + \theta^2 X_{t-2} - \cdots
$$

$$
\text{B} \qquad Z_t = X_t - \frac{1}{\theta} X_{t-1} + \frac{1}{\theta^2} X_{t-2} - \cdots
$$

If $|\theta| < 1$, the series for A converges whereas that for B does not. Thus an estimation procedure which involves estimating the residuals (see Section 4.3.1) will lead naturally to model A. Thus if $|\theta| < 1$, model A is said to be invertible whereas model B is not. The imposition of the invertibility condition ensures that there is a unique MA process for a given ac.f.

The invertibility condition for the general-order MA process is best expressed by using the backward shift operator, denoted by B, which is defined by

$$
B^j X_t = X_{t-j} \qquad \text{for all } j
$$

Then equation (3.2) may be written as

$$
X_t = (\beta_0 + \beta_1 B + \cdots + \beta_q B^q) Z_t
$$
$$
= \theta(B) Z_t
$$

where $\theta(B)$ is a polynomial of order q in B. An MA process of order q is invertible if the roots of the equation (regarding B as a complex variable and not an operator)

$$\theta(B) = \beta_0 + \beta_1 B + \cdots + \beta_q B^q = 0$$

all lie outside the unit circle (Box and Jenkins, 1970, p. 50). For example, in the first-order case we have $\theta(B) = 1 + \theta B$, which has root $B = -1/\theta$. Thus the root is outside the unit circle provided that $|\theta| < 1$.

MA processes have been used in many areas, particularly econometrics. For example economic indicators are affected by a variety of 'random' events such as strikes, government decisions, shortages of key materials and so on. Such events will not only have an immediate effect but may also affect economic indicators to a lesser extent in several subsequent periods, and so it is at least plausible that an MA process may be appropriate.

Note that an arbitrary constant, μ say, may be added to the right-hand side of equation (3.2) to give a process with mean μ. This does not affect the ac.f. and has been omitted for simplicity.

3.4.4 Autoregressive processes

Suppose that $\{Z_t\}$ is a purely random process with mean zero and variance σ_Z^2. Then a process $\{X_t\}$ is said to be an autoregressive process of order p if

$$X_t = \alpha_1 X_{t-1} + \cdots + \alpha_p X_{t-p} + Z_t \qquad (3.3)$$

This is rather like a multiple regression model, but X_t is regressed not on independent variables but on past values of X_t; hence the prefix 'auto'. An autoregressive process of order p will be abbreviated to an AR(p) process.

(a) *First-order process*

For simplicity, we begin by examining the first-order case, where $p = 1$. Then

$$X_t = \alpha X_{t-1} + Z_t \qquad (3.4)$$

The AR(1) process is sometimes called the Markov process, after the Russian A. A. Markov. By successive substitution in (3.4) we may write

$$X_t = \alpha(\alpha X_{t-2} + Z_{t-1}) + Z_t$$
$$= \alpha^2(\alpha X_{t-3} + Z_{t-2}) + \alpha Z_{t-1} + Z_t$$

and eventually we find that X_t may be expressed as an infinite-order MA process in the form (provided $-1 < \alpha < +1$)

$$X_t = Z_t + \alpha Z_{t-1} + \alpha^2 Z_{t-2} + \cdots$$

This duality between AR and MA processes is useful for a variety of purposes. Rather than use successive substitution, it is simpler to use the backward shift operator B. Then equation (3.4) may be written

$$(1 - \alpha B)X_t = Z_t$$

so that

$$X_t = Z_t/(1 - \alpha B)$$
$$= (1 + \alpha B + \alpha^2 B^2 + \cdots)Z_t$$
$$= Z_t + \alpha Z_{t-1} + \alpha^2 Z_{t-2} + \cdots$$

When expressed in this form it is clear that

$$E(X_t) = 0$$
$$\text{Var}(X_t) = \sigma_Z^2(1 + \alpha^2 + \alpha^4 + \cdots)$$

Thus the variance is finite provided that $|\alpha| < 1$, in which case

$$\text{Var}(X_t) = \sigma_X^2 = \sigma_Z^2/(1 - \alpha^2)$$

The acv.f. is given by

$$\gamma(k) = E[X_t X_{t+k}]$$
$$= E[(\Sigma \alpha^i Z_{t-i})(\Sigma \alpha^j Z_{t+k-j})]$$
$$= \sigma_Z^2 \sum_{i=0}^{\infty} \alpha^i \alpha^{k+i} \qquad \text{for } k \geqslant 0$$

which converges for $|\alpha| < 1$ to

$$\gamma(k) = \alpha^k \sigma_Z^2/(1 - \alpha^2)$$
$$= \alpha^k \sigma_X^2$$

For $k < 0$, we find $\gamma(k) = \gamma(-k)$. Since $\gamma(k)$ does not depend on t, an AR process of order 1 is second-order stationary provided that $|\alpha| < 1$. The ac.f. is given by

$$\rho(k) = \alpha^k \qquad k = 0, 1, 2, \ldots$$

To get an even function defined for all integer k we can write

$$\rho(k) = \alpha^{|k|} \qquad k = 0, \pm 1, \pm 2, \ldots$$

The ac.f. may also be obtained more simply by assuming *a priori* that the process is stationary, in which case $E(X_t)$ must be zero. Multiply through equation (3.4) by X_{t-k} (**not** X_{t+k}!) and take expectations. Then we find, for $k > 0$, that

$$\gamma(-k) = \alpha \gamma(-k+1)$$

assuming that $E(Z_t X_{t-k}) = 0$ for $k > 0$. Since $\gamma(k)$ is an even function, we must also have

$$\gamma(k) = \alpha \gamma(k-1) \qquad \text{for } k > 0$$

Now $\gamma(0) = \sigma_X^2$, and so $\gamma(k) = \alpha^k \sigma_X^2$ for $k \geqslant 0$. Thus $\rho(k) = \alpha^k$ for $k \geqslant 0$. Now since

$|\rho(k)| \leqslant 1$, we must have $|\alpha| \leqslant 1$. But if $|\alpha| = 1$, then $|\rho(k)| = 1$ for all k, which is a degenerate case. Thus $|\alpha| < 1$ for a proper stationary process.

The above method of obtaining the ac.f. is often used, even though it involves 'cheating' a little by making an initial assumption of stationarity.

Three examples of the ac.f. of a first-order AR process are shown in Figure 3.1 for $\alpha = 0.8$, 0.3, -0.8. Note how quickly the ac.f. decays when $\alpha = 0.3$, and note how the ac.f. alternates when α is negative.

(b) *General-order case*

As in the first-order case, we can express an AR process of finite order as an MA process of infinite order. This may be done by successive substitution, or by using the backward shift operator. Then equation (3.3) may be written as

$$(1 - \alpha_1 B - \cdots - \alpha_p B^p)X_t = Z_t$$

or

$$X_t = Z_t/(1 - \alpha_1 B - \cdots - \alpha_p B^p)$$
$$= f(B)Z_t$$

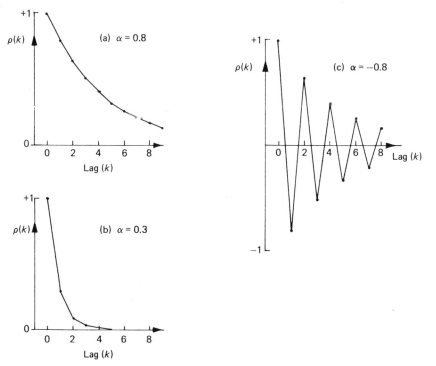

Figure 3.1   Three examples of the autocorrelation function of a first-order autoregressive process with, (a) $\alpha = 0.8$; (b) $\alpha = 0.3$; (c) $\alpha = -0.8$.

where

$$f(B) = (1 - \alpha_1 B - \cdots - \alpha_p B^p)^{-1}$$
$$= (1 + \beta_1 B + \beta_2 B^2 + \cdots)$$

The relationship between the αs and the βs may then be found. Having expressed X_t as an MA process, it follows that $E(X_t) = 0$. The variance is finite provided that $\Sigma \beta_i^2$ converges, and this is a necessary condition for stationarity. The acv.f. is given by

$$\gamma(k) = \sigma_Z^2 \sum_{i=0}^{\infty} \beta_i \beta_{i+k} \qquad \text{where } \beta_0 = 1$$

A sufficient condition for this to converge, and hence for stationarity, is that $\Sigma |\beta_i|$ converges.

We can in principle find the ac.f. of the general-order AR process using the above procedure, but the $\{\beta_i\}$ may be algebraically hard to find. The alternative simpler way is to **assume** the process is stationary, multiply through equation (3.3) by X_{t-k}, take expectations, and divide by σ_X^2, assuming that the variance of X_t is finite. Then, using the fact that $\rho(k) = \rho(-k)$ for all k, we find

$$\rho(k) = \alpha_1 \rho(k-1) + \cdots + \alpha_p \rho(k-p) \qquad \text{for all } k > 0$$

This set of equations is called the Yule-Walker equations after G. U. Yule and Sir Gilbert Walker. It is a set of difference equations and has the general solution

$$\rho(k) = A_1 \pi_1^{|k|} + \cdots + A_p \pi_p^{|k|}$$

where $\{\pi_i\}$ are the roots of the so-called auxiliary equation

$$y^p - \alpha_1 y^{p-1} \cdots - \alpha_p = 0$$

The constants $\{A_i\}$ are chosen to satisfy the initial conditions depending on $\rho(0) = 1$, which means that $\Sigma A_i = 1$. The first $(p-1)$ Yule-Walker equations provide $(p-1)$ further restrictions on the $\{A_i\}$ using $\rho(0) = 1$ and $\rho(k) = \rho(-k)$.

From the general form of $\rho(k)$, it is clear that $\rho(k)$ tends to zero as k increases provided that $|\pi_i| < 1$ for all i, and this is a necessary and sufficient condition for the process to be stationary.

An equivalent way of expressing the stationarity condition is to say that the roots of the equation

$$\phi(B) = 1 - \alpha_1 B - \cdots - \alpha_p B^p = 0 \qquad (3.5)$$

must lie outside the unit circle (Box and Jenkins, 1970, Section 3.2).

Of particular interest is the AR(2) process, when π_1, π_2 are the roots of the quadratic equation

$$y^2 - \alpha_1 y - \alpha_2 = 0$$

Thus $|\pi_i| < 1$ if

$$\left| \frac{\alpha_1 \pm \sqrt{(\alpha_1^2 + 4\alpha_2)}}{2} \right| < 1$$

from which it can be shown (Exercise 3.6) that the stationarity region is the triangular region satisfying

$$\alpha_1 + \alpha_2 < 1$$
$$\alpha_1 - \alpha_2 > -1$$
$$\alpha_2 > -1$$

The roots are real if $\alpha_1^2 + 4\alpha_2 > 0$, in which case the ac.f. decreases exponentially with k, but the roots are complex if $\alpha_1^2 + 4\alpha_2 < 0$, in which case we find that the ac.f. is a damped cosine wave. (See Example 3.1 at the end of this section.)

When the roots are real, the constants A_1, A_2 are found as follows. Since $\rho(0) = 1$, we have

$$A_1 + A_2 = 1$$

From the first of the Yule-Walker equations, we have

$$\rho(1) = \alpha_1 \rho(0) + \alpha_2 \rho(-1)$$
$$= \alpha_1 + \alpha_2 \rho(1)$$

Thus

$$\rho(1) = \alpha_1 / (1 - \alpha_2)$$
$$= A_1 \pi_1 + A_2 \pi_2$$
$$= A_1 \pi_1 + (1 - A_1) \pi_2$$

Hence we find

$$A_1 = [\alpha_1 / (1 - \alpha_2) - \pi_2] / (\pi_1 - \pi_2)$$
$$A_2 = 1 - A_1$$

AR processes have been applied to many situations in which it is reasonable to assume that the present value of a time series depends on the immediate past values together with a random error. For simplicity we have only considered processes with mean zero, but non-zero means may be dealt with by rewriting equation (3.3) in the form

$$X_t - \mu = \alpha_1 (X_{t-1} - \mu) + \cdots + \alpha_p (X_{t-p} - \mu) + Z_t$$

This does not affect the ac.f.

Example 3.1   Consider the AR(2) process given by

$$X_t = X_{t-1} - \tfrac{1}{2}X_{t-2} + Z_t$$

Is this process stationary? If so, what is its ac.f.?

In order to answer the first question we find the roots of equation (3.5), which in this case is

$$\phi(B) = 1 - B + \tfrac{1}{2}B^2 = 0$$

The roots of this equation (regarding B as a variable) are $1 \pm i$. As the modulus of both roots exceeds one, the roots are both outside the unit circle and so the process is stationary.

In order to find the ac.f. of the process, we use the first Yule-Walker equation to give

$$\rho(1) = \rho(0) - \tfrac{1}{2}\rho(-1)$$
$$= 1 - \tfrac{1}{2}\rho(1)$$

giving $\rho(1) = 2/3$.

For $k \geqslant 2$, the Yule-Walker equations are

$$\rho(k) = \rho(k-1) - \tfrac{1}{2}\rho(k-2)$$

We could find $\rho(2)$, then $\rho(3)$, and so on by successive substitution, but it is easier to find the general solution by solving as a difference equation, which has the auxiliary equation

$$y^2 - y + \tfrac{1}{2} = 0$$

with roots $y = (1 \pm i)/2 = [\cos(\pi/4) \pm i\,\sin(\pi/4)]/\sqrt{2} = e^{\pm i\pi/4}/\sqrt{2}$. Since $\alpha_1^2 + 4\alpha_2 = (1-2)$ is less than zero, the ac.f. is a damped cosine wave. Using $\rho(0) = 1$ and $\rho(1) = 2/3$, some messy trigonometry and algebra gives

$$\rho_k = \left(\frac{1}{\sqrt{2}}\right)^k \left(\cos\frac{\pi k}{4} + \frac{1}{3}\sin\frac{\pi k}{4}\right)$$

for $k = 0, 1, 2, \ldots$.

3.4.5 Mixed ARMA models

A useful class of models for time series is formed by combining MA and AR processes. A mixed autoregressive/moving-average process containing p AR terms and q MA terms is said to be an ARMA process of order (p, q). It is given by

$$X_t = \alpha_1 X_{t-1} + \cdots + \alpha_p X_{t-p} + Z_t + \beta_1 Z_{t-1}$$
$$+ \cdots + \beta_q Z_{t-q} \tag{3.6}$$

Using the backward shift operator B, equation (3.6) may be written in the form

$$\phi(B)X_t = \theta(B)Z_t \qquad (3.6a)$$

where $\phi(B)$, $\theta(B)$ are polynomials of order p, q respectively, such that

$$\phi(B) = 1 - \alpha_1 B - \cdots - \alpha_p B^p$$

and

$$\theta(B) = 1 + \beta_1 B + \cdots + \beta_q B^q$$

As for an AR process, the values of $\{\alpha_i\}$ which make the process stationary are such that the roots of

$$\phi(B) = 0$$

lie outside the unit circle. As for an MA process, the values of $\{\beta_i\}$ which make the process invertible are such that the roots of

$$\theta(B) = 0$$

lie outside the unit circle.

It is straightforward in principle, though algebraically rather tedious, to calculate the ac.f. of an ARMA process, but this will not be discussed here. (See Exercise 3.11; and see Box and Jenkins, 1970, Section 3.4).

The importance of ARMA processes lies in the fact that a stationary time series may often be described by an ARMA model involving fewer parameters than a pure MA or AR process by itself.

It is sometimes helpful to express an ARMA model as a pure MA process in the form

$$X_t = \psi(B)Z_t \qquad (3.6b)$$

where $\psi(B) = \Sigma \psi_i B^i$ is the MA operator which may be of infinite order. The ψ weights, $\{\psi_i\}$, can be useful in calculating forecasts (see Chapter 5) and in assessing the properties of a model (e.g. see Exercise 3.11). By comparison with equation (3.6a), we see that $\psi(B) = \theta(B)/\phi(B)$. Alternatively, it can be helpful to express an ARMA model as a pure AR process in the form

$$\pi(B)X_t = Z_t \qquad (3.6c)$$

where $\pi(B) = \phi(B)/\theta(B)$. By convention we write $\pi(B) = 1 - \sum_{i \geq 1} \pi_i B^i$, since the natural way to write an AR model is in the form

$$X_t = \sum_{i=1}^{\infty} \pi_i X_{t-i} + Z_t$$

By comparing (3.6b) and (3.6c), we see that

$$\pi(B)\psi(B) = 1$$

The ψ weights or π weights may be obtained directly by division or by equating powers of B in an equation such as

$$\psi(B)\phi(B) = \theta(B)$$

Example 3.2 Find the ψ weights and π weights for the ARMA(1, 1) process given by

$$X_t = 0.5X_{t-1} + Z_t - 0.3Z_{t-1}$$

Here $\phi(B) = (1 - 0.5B)$ and $\theta(B) = (1 - 0.3B)$, so the process is stationary and invertible. Then

$$\psi(B) = \theta(B)/\phi(B) = (1 - 0.3B)(1 - 0.5B)^{-1}$$
$$= (1 - 0.3B)(1 + 0.5B + 0.5^2 B^2 + \cdots)$$
$$= 1 + 0.2B + 0.1B^2 + 0.005B^3 + \cdots$$

Hence

$$\psi_i = 0.2 \times 0.5^{i-1} \qquad \text{for } i = 1, 2, \ldots$$

Similarly we find

$$\pi_i = 0.2 \times 0.3^{i-1} \qquad \text{for } i = 1, 2, \ldots$$

Note that both the ψ weights and π weights die away quickly, and this also indicates a stationary, invertible process.

3.4.6 Integrated ARIMA models

In practice most time series are non-stationary. In order to fit a stationary model, such as those discussed in Sections 3.4.3–3.4.5, it is necessary to remove non-stationary sources of variation. If the observed time series is non-stationary in the mean then we can difference the series, as suggested in Section 2.5.3, and this approach is widely used in econometrics. If X_t is replaced by $\nabla^d X_t$ in equation (3.6) then we have a model capable of describing certain types of non-stationary series. Such a model is called an 'integrated' model because the stationary model which is fitted to the differenced data has to be summed or 'integrated' to provide a model for the non-stationary data. Writing

$$W_t = \nabla^d X_t = (1 - B)^d X_t$$

the general autoregressive integrated moving average process (abbreviated ARIMA process) is of the form

$$W_t = \alpha_1 W_{t-1} + \cdots + \alpha_p W_{t-p} + Z_t + \cdots + \beta_q Z_{t-q} \qquad (3.7)$$

By analogy with equation (3.6a), we may write equation (3.7) in the form

$$\phi(B)W_t = \theta(B)Z_t \tag{3.7a}$$

or

$$\phi(B)(1-B)^d X_t = \theta(B)Z_t \tag{3.7b}$$

Thus we have an ARMA(p, q) model for W_t, while the model in equation (3.7b), describing the dth differences of X_t, is said to be an ARIMA process of order (p, d, q). The model for X_t is clearly non-stationary, as the AR operator $\phi(B)(1-B)^d$ has d roots on the unit circle. In practice the value of d is often taken to be one. Note that the random walk can be regarded as an ARIMA$(0, 1, 0)$ process.

ARIMA models can be generalized to include seasonal terms, as discussed in Section 4.6.

3.4.7 The general linear process

The infinite-order MA process with non-zero mean given by

$$X_t - \mu = \sum_{i=0}^{\infty} \beta_i Z_{t-i} \tag{3.8}$$

is sometimes called the general linear process since processes of this type can be obtained by passing a purely random process through a linear system (see Chapter 9). Both MA and AR processes are special cases of the general linear process and the duality between these two classes is easily demonstrated using equations (3.6b) and (3.6c).

†3.4.8 Continuous process

So far, we have only considered stochastic processes in discrete time, because these are the main type of process the statistician uses in practice. Continuous-time processes have been used in some applications, notably in the study of control theory by electrical engineers. Here we shall only indicate some of the problems connected with their use.

By analogy with a discrete-time purely random process, we might expect to define a continuous-time purely random process as having an ac.f. given by

$$\rho(\tau) = \begin{cases} 1 & \tau = 0 \\ 0 & \tau \neq 0 \end{cases}$$

However, this is a discontinuous function, and it can be shown that such a process would have an infinite variance and hence be a physically unrealizable

†This section should be omitted at first reading.

phenomenon. Nevertheless, some processes which arise in practice do appear to have the properties of continuous-time white noise even when sampled at quite small discrete intervals. We may approximate continuous-time white noise by considering a purely random process in discrete time at intervals Δt, and letting $\Delta t \to 0$, or by considering a process in continuous time with ac.f. $\rho(\tau) = e^{-\lambda|\tau|}$ and letting $\lambda \to \infty$ so that the ac.f. decays very quickly.

As an example of the difficulties involved with continuous-time processes, we briefly consider a first-order continuous AR process. A first-order discrete AR process may be written in terms of X_t, ∇X_t and Z_t. As differencing in discrete time corresponds to differentiation in continuous time, a natural way of trying to define a continuous first-order AR process is by

$$\frac{dX(t)}{dt} + aX(t) = Z(t) \qquad (3.9)$$

where a is a constant, and $Z(t)$ denotes continuous white noise. In the theory of Brownian motion, this is called Langevin's equation. However, as $Z(t)$ does not physically exist, it is more legitimate to write equation (3.9) in a form involving infinitesimal small changes as

$$dX(t) + aX(t)\, dt = dU(t) \qquad (3.10)$$

where $\{U(t)\}$ is a process with orthogonal increments such that the random variables $[U(t_2) - U(t_1)]$ and $[U(t_4) - U(t_3)]$ are uncorrelated for any two non-overlapping intervals (t_1, t_2) and (t_3, t_4). It can then be shown that the process $X(t)$ defined in equation (3.10) has ac.f.

$$\rho(\tau) = e^{-a|\tau|}$$

which is similar to the ac.f. of a first-order discrete AR process in that both decay exponentially. However, the rigorous study of continuous processes, such as that in equation (3.9), requires considerable mathematical machinery, including a knowledge of stochastic integration, and we will not pursue it here. The reader is referred for example to Yaglom (1962) and Cox and Miller (1968, Section 7.4).

†3.5   THE WOLD DECOMPOSITION THEOREM

This section gives a brief introduction to a famous result, called the Wold decomposition theorem, which is of mainly theoretical interest. The treatment in this section is a shortened version of that given by Cox and Miller (1968). The Wold decomposition theorem says that any discrete stationary process can be expressed as the sum of two uncorrelated processes, one purely deterministic and one purely indeterministic. The terms 'deterministic' and

---

†This section should be omitted at first reading.

'indeterministic' are defined as follows. We can regress X_t on $(X_{t-q}, X_{t-q-1}, \ldots)$ and denote the residual variance from the resulting linear regression model by τ_q^2. As $\tau_q^2 \leqslant \mathrm{Var}(X_t)$, it is clear that, as q increases, τ_q^2 is a non-decreasing bounded sequence and therefore tends to a limit as $q \to \infty$. If $\lim_{q \to \infty} \tau_q^2 = \mathrm{Var}(X_t)$ then linear regression on the remote past is useless for prediction purposes, and we say that $\{X_t\}$ is **purely indeterministic**. But if $\lim_{q \to \infty} \tau_q^2$ is zero then the process can be forecast exactly, and we say that $\{X_t\}$ is purely **deterministic**.

All the stationary processes we have considered in this chapter, such as AR and MA processes, are purely indeterministic. The best-known examples of purely deterministic processes are sinusoidal processes (see Exercise 3.14), such as

$$X_t = g \cos(\omega t + \theta) \tag{3.11}$$

where g is a constant, ω is a constant in $(0, \pi)$ called the frequency of the process, and θ is a random variable called the phase which is uniformly distributed on $(0, 2\pi)$ but which is fixed for a single realization. Note that we must include the term θ so that

$$E(X_t) = 0 \qquad \text{for all } t$$

otherwise (3.11) would not define a stationary process. As θ is fixed for a single realization, once enough values of X_t have been observed to evaluate θ, all subsequent values of X_t are completely determined. It is then obvious that (3.11) defines a purely deterministic process.

The Wold decomposition theorem also says that the purely indeterministic component can be written as the linear sum of an 'innovation' process $\{Z_t\}$, which is a sequence of uncorrelated random variables. A special class of processes of particular interest arise when the Zs are independent and not merely uncorrelated, as we then have a general linear process (Section 3.4.7). On the other hand when processes are generated in a non-linear way the Wold decomposition is usually of little interest.

The concept of a purely indeterministic process is a useful one, and most of the stationary stochastic processes which are considered in the rest of this book are of this type.

EXERCISES

In all the following questions $\{Z_t\}$ is a discrete, purely random process, such that $E(Z_t)=0$, $\mathrm{Var}(Z_t)=\sigma_Z^2$, $\mathrm{Cov}(Z_t, Z_{t+k})=0$, $k \neq 0$.
Exercise 3.14 is harder than the others and may be omitted.

3.1   Find the ac.f. of the second-order MA process given by

$$X_t = Z_t + 0.7Z_{t-1} - 0.2Z_{t-2}$$

3.2   Show that the ac.f. of the mth-order MA process given by

$$X_t = \sum_{k=0}^{m} Z_{t-k}/(m+1)$$

is

$$\rho(k) = \begin{cases} (m+1-k)/(m+1) & k=0, 1, \ldots, m \\ 0 & k>m \end{cases}$$

3.3   Show that the infinite-order MA process $\{X_t\}$ defined by

$$X_t = Z_t + C(Z_{t-1} + Z_{t-2} + \cdots)$$

where C is a constant, is non-stationary. Also show that the series of first differences $\{Y_t\}$ defined by

$$Y_t = X_t - X_{t-1}$$

is a first-order MA process and is stationary. Find the ac.f. of $\{Y_t\}$.

3.4   Find the ac.f. of the first-order AR process defined by

$$X_t - \mu = 0.7(X_{t-1} - \mu) + Z_t$$

Plot $\rho(k)$ for $k = -6, -5, \ldots, -1, 0, +1, \ldots, +6$.

3.5   If $X_t = \mu + Z_t + \beta Z_{t-1}$, where μ is a constant, show that the ac.f. does not depend on μ.

3.6   Find the values of λ_1, λ_2, such that the second-order AR process defined by

$$X_t = \lambda_1 X_{t-1} + \lambda_2 X_{t-2} + Z_t$$

is stationary. If $\lambda_1 = 1/3$, $\lambda_2 = 2/9$, show that the ac.f. of X_t is given by

$$\rho(k) = \frac{16}{21}\left(\frac{2}{3}\right)^{|k|} + \frac{5}{21}\left(-\frac{1}{3}\right)^{|k|} \qquad k=0, \pm 1, \pm 2, \ldots.$$

3.7.   Explain what is meant by a weakly (or second-order) stationary process, and define the ac.f. $\rho(u)$ for such a process. Show that $\rho(u) = \rho(-u)$ and that $|\rho(u)| \leqslant 1$.

Show that the ac.f. of the stationary second-order AR process

$$X_t = \frac{1}{12} X_{t-1} + \frac{1}{12} X_{t-2} + Z_t$$

is given by

$$\rho(k) = \frac{45}{77}\left(\frac{1}{3}\right)^{|k|} + \frac{32}{77}\left(-\frac{1}{4}\right)^{|k|} \qquad k=0, \pm 1, \pm 2, \ldots$$

3.8   The stationary process $\{X_t\}$ has acv.f. $\gamma_X(k)$. A new stationary process $\{Y_t\}$ is defined by $Y_t = X_t - X_{t-1}$. Obtain the acv.f. of $\{Y_t\}$ in terms of $\gamma_X(k)$ and find $\gamma_Y(k)$ when $\gamma_X(k) = \lambda^{|k|}$.

3.9   For each of the following models:

(a) $X_t = 0.3X_{t-1} + Z_t$
(b) $X_t = Z_t - 1.3Z_{t-1} + 0.4Z_{t-2}$
(c) $X_t = 0.5X_{t-1} + Z_t - 1.3Z_{t-1} + 0.4Z_{t-2}$

express the model in B notation and determine whether the model is stationary and/or invertible. For model (a) find the equivalent MA representation.

3.10 A stationary process $\{X_t\}$ can be represented in the form $X_t = \psi(B)Z_t$ or $\pi(B)X_t = Z_t$. The autocovariance generating function is given by

$$\Gamma(s) = \sum_{k=-\infty}^{\infty} \gamma_X(k)s^k$$

Show that $\Gamma(s) = \sigma_Z^2 \psi(s)\psi(1/s) = \sigma_Z^2/[\pi(s)\pi(1/s)]$.
(Hint: Equate coefficients of s^k.)

3.11 Show that the ac.f. of the ARMA(1, 1) model

$$X_t = \alpha X_{t-1} + Z_t + \beta Z_{t-1}$$

is given by

$$\rho(1) = (1 + \alpha\beta)(\alpha + \beta)/(1 + \beta^2 + 2\alpha\beta)$$
$$\rho(k) = \alpha\rho(k-1) \qquad k = 2, 3, \ldots$$

3.12 For the model $(1 - B)(1 - 0.2B)X_t = (1 - 0.5B)Z_t$:

(a) Classify the model as an ARIMA(p, d, q) process (i.e. find p, d, q).
(b) Determine whether the process is stationary.
(c) Evaluate the first three ψ weights.
(d) Evaluate the first four π weights.

3.13 Show that the AR(2) process

$$X_t = X_{t-1} + cX_{t-2} + Z_t$$

is stationary provided $-1 < c < 0$. Find the autocorrelation function when $c = -3/16$.
      Show that the AR(3) process

$$X_t = X_{t-1} + cX_{t-2} - cX_{t-3} + Z_t$$

is non-stationary for all values of c.

3.14 For a complex-valued process $X(t)$, with (complex) mean μ, the acv.f. is defined by

$$\gamma(\tau) = E\{[X(t) - \mu][\bar{X}(t+\tau) - \bar{\mu}]\}$$

where the overbar denotes the complex conjugate. Show that the process $X(t) = Y e^{i\omega t}$ is second-order stationary, where Y is a complex random variable mean zero which does not depend on t, and ω is a real constant. One useful form for the random variable Y occurs when it takes the form $g e^{i\theta}$, where g is a constant and θ is a uniformly distributed random variable on $(0, 2\pi)$. Show that $E(Y) = 0$ in this case (see Yaglom, 1962, Section 2.8; but note that the autocovariance function is called the correlation function by Yaglom).

4

Estimation in the time domain

In Chapter 3 we introduced several different types of probability model which may be used to describe time series. In this chapter we discuss the problem of fitting a suitable model to an observed time series, confining ourselves to the discrete-time case. The major diagnostic tool in this chapter is the sample autocorrelation function. Inference based on this function is often called an analysis in the **time domain**.

4.1   ESTIMATING THE AUTOCOVARIANCE AND AUTOCORRELATION FUNCTIONS

We have already noted in Section 3.3 that the theoretical ac.f. is an important tool for describing the properties of a stationary stochastic process. In Section 2.7 we heuristically introduced the sample ac.f. of an observed time series, and this is an intuitively reasonable estimate of the theoretical ac.f., provided the series is stationary. This section investigates the properties of the sample ac.f. more closely.

Let us look first at the autocovariance function (acv.f.). The sample autocovariance coefficient at lag k (see equation (2.7)), given by

$$c_k = \sum_{t=1}^{N-k} (x_t - \bar{x})(x_{t+k} - \bar{x})/N \qquad (4.1)$$

is the usual estimator for the theoretical autocovariance coefficient $\gamma(k)$ at lag k. The properties of this estimator are discussed by Jenkins and Watts (1968, Section 5.3.3) and Priestley (1981, Chapter 5). It can be shown that the bias in c_k is of order $1/N$. However,

$$\lim_{N \to \infty} E(c_k) = \gamma(k)$$

so that the estimator is asymptotically unbiased.

It can also be shown that

$$\text{Cov}(c_k, c_m) \simeq \sum_{r=-\infty}^{\infty} \{\gamma(r)\gamma(r+m-k)+\gamma(r+m)\gamma(r-k)\}/N \qquad (4.2)$$

When $m=k$, formula (4.2) gives us the variance of c_k and hence the mean square error of c_k. Formula (4.2) also highlights the fact that successive values of c_k may be highly correlated and this increases the difficulty of interpreting the correlogram.

Jenkins and Watts (1968, Chapter 5) compare the estimator (4.1) with the alternative estimator

$$c_k' = \sum_{t=1}^{N-k} (x_t - \bar{x})(x_{t+k} - \bar{x})/(N-k)$$

This is used by some authors because it has a smaller bias, but Jenkins and Watts conjecture that it generally has a higher mean square error. In any case it is the biased estimator in equation (4.1) which gives a function having a useful property called positive semi-definiteness (Priestley, 1981), which leads to a finite Fourier transform which is non-negative. The latter is useful in estimating the spectrum (see Chapter 7).

A third method of estimating the acv.f. is to use Quenouille's method of bias reduction, otherwise known as **jackknife** estimation. In this procedure the time series is divided into two halves, and the sample acv.f. is estimated from each half of the series and also from the whole series. If the three resulting estimates of $\gamma(k)$ are denoted by c_{k1}, c_{k2} and c_k in an obvious notation, then the jackknife estimate is given by

$$\tilde{c}_k = 2c_k - \tfrac{1}{2}(c_{k1} + c_{k2}) \qquad (4.3)$$

It can be shown that this estimator reduces the bias from order $1/N$ to order $1/N^2$. It has an extra advantage in that one can see if both halves of the time series have similar properties and hence see if the time series is stationary. However, the method has the disadvantage that it requires extra computation. It is also sensitive to non-stationarity in the mean, and c_{k1}, c_{k2} should be compared with the overall c_k as well as with each other.

Having estimated the acv.f., we then take

$$r_k = c_k/c_0 \qquad (4.4)$$

as an estimator for $\rho(k)$. The properties of r_k are rather more difficult to find than those of c_k because it is the ratio of two random variables. It can be shown that r_k is generally biased. The bias can be reduced by jackknifing as in equation (4.3). The jackknife estimator is given in an obvious notation by

$$\tilde{r}_k = 2r_k - \tfrac{1}{2}(r_{k1} + r_{k2})$$

A general formula for the variance of r_k is given by Kendall, Stuart and Ord (1983, Section 48.1) and depends on **all** the autocorrelation coefficients of the process. We will only consider the properties of r_k when sampling from a purely random process, when all the theoretical autocorrelation coefficients are zero except at lag zero. These results help us to decide if the observed values of r_k from a given time series are significantly different from zero.

Suppose that $x_1, \ldots, x_N$ are independent and identically distributed random variables with arbitrary mean. Then it can be shown (Kendall, Stuart and Ord, 1983, Chapter 48) that

$$E(r_k) \simeq -1/N$$

$$\text{Var}(r_k) \simeq 1/N$$

and that r_k is asymptotically normally distributed under weak conditions. Thus having plotted the correlogram, as described in Section 2.7, we can plot approximate 95% confidence limits at $-1/N \pm 2/\sqrt{N}$, which are often further approximated to $\pm 2/\sqrt{N}$. Observed values of r_k which fall outside these limits are 'significantly' different from zero at the 5% level. However, when interpreting a correlogram, it must be remembered that the overall probability of getting a coefficient outside these limits, given that the data really are random, increases with the number of coefficients plotted. For example if the first 20 values of r_k are plotted then one expects one 'significant' value on average even if the data really are random. Thus, if only one or two coefficients are 'significant', the size and lag of these coefficients must be taken into account when deciding if a set of data is random. Values well outside the 'null' confidence limits indicate non randomness. So also does a significant coefficient at a lag which has some physical interpretation, such as lag 1 or a lag corresponding to seasonal variation.

Figure 4.1 shows the correlogram for 100 observations, generated on a computer, which are supposed to be independent normally distributed variables. The confidence limits are approximately $\pm 2/\sqrt{100} = \pm 0.2$. We see that 2 of the first 20 values of r_k lie just outside the significance limits. As these occur at apparently arbitrary lags (namely 12 and 17) we conclude that there is no firm evidence to reject the hypothesis that the observations are independently distributed.

4.1.1   Interpreting the correlogram

We have already given some general advice on interpreting correlograms in Section 2.7.2. The correlogram is also helpful in identifying which type of ARIMA model gives the best representation of an observed time series. A correlogram like that in Figure 2.3, where the values of r_k do not come down to zero reasonably quickly, indicates non-stationarity and so the series needs to be differenced. For stationary series, the correlogram is compared with the

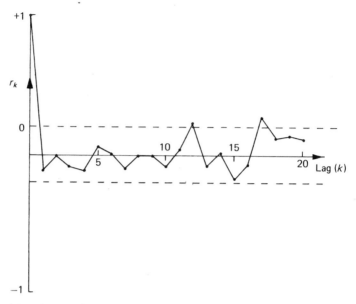

Figure 4.1   The correlogram of 100 'independent' normally distributed observations.

theoretical ac.f.s of different ARMA processes in order to choose the one which is most appropriate. The ac.f. of an MA(q) process is easy to recognize as it 'cuts off' at lag q, whereas the ac.f. of an AR(p) process is a mixture of damped exponentials and sinusoids and dies out slowly (or attenuates). The ac.f. of a mixed ARMA model will also generally attenuate rather than 'cut off'. For example, suppose we find that r_1 is significantly different from zero but that subsequent values of r_k are all close to zero. Then an MA(1) model is indicated since its theoretical ac.f. is of this form. Alternatively, if $r_1, r_2, r_3, \ldots$ appear to be decreasing exponentially, then an AR(1) model may be appropriate.

The interpretation of correlograms is one of the hardest aspects of time-series analysis and practical experience is a 'must'. Inspection of the partial autocorrelation function (see Section 4.2.2) can provide some help.

†4.1.2   Ergodic theorems

The fact that one can obtain consistent estimates of the properties of a stationary process from a single finite realization is not immediately obvious. However some theorems, called ergodic theorems, have been proved which show that, for most stationary processes which are likely to be met in practice, the sample moments of an observed record of length T converge (in mean

†This section may be omitted at a first reading.

square) to the corresponding population moments as $T \to \infty$. In other words, time averages for a single realization converge to ensemble averages. See for example Yaglom (1962, Section 1.4). We will not pursue the topic here.

4.2 FITTING AN AUTOREGRESSIVE PROCESS

Having estimated the ac.f. of a given time series, we should have some idea as to which stochastic process will provide a suitable model. If an AR process is thought to be appropriate, there are two related questions:

(a) What is the order of the process?
(b) How can we estimate the parameters of the process?

We will consider question (b) first.

4.2.1 Estimating the parameters of an autoregressive process

Suppose we have an AR process of order p, with mean μ, given by

$$X_t - \mu = \alpha_1(X_{t-1} - \mu) + \cdots + \alpha_p(X_{t-p} - \mu) + Z_t \qquad (4.5)$$

Given N observations $x_1, \ldots, x_N$, the parameters $\mu, \alpha_1, \ldots, \alpha_p$ may be estimated by least squares by minimizing

$$S = \sum_{t=p+1}^{N} [x_t - \mu - \alpha_1(x_{t-1} - \mu) - \cdots - \alpha_p(x_{t-p} - \mu)]^2$$

with respect to $\mu, \alpha_1, \ldots, \alpha_n$. If the Z_t process is normal, then the least squares estimates are in addition maximum likelihood estimates (Jenkins and Watts, 1968, Section 5.1) conditional on the first p values in the time series being fixed.

In the first-order case, with $p = 1$, we find (see Exercise 4.1)

$$\hat{\mu} = \frac{\bar{x}_{(2)} - \hat{\alpha}_1 \bar{x}_{(1)}}{1 - \hat{\alpha}_1} \qquad (4.6)$$

and

$$\hat{\alpha}_1 = \frac{\sum_{t=1}^{N-1} (x_t - \hat{\mu})(x_{t+1} - \hat{\mu})}{\sum_{t=1}^{N-1} (x_t - \hat{\mu})^2} \qquad (4.7)$$

where $\bar{x}_{(1)}, \bar{x}_{(2)}$ are the means of the first and last $(N-1)$ observations. Now since

$$\bar{x}_{(1)}^* \simeq \bar{x}_{(2)} \simeq \bar{x}$$

we have approximately that

$$\hat{\mu} = \bar{x} \tag{4.8}$$

This approximate estimator, which is intuitively appealing, is nearly always used in preference to (4.6). Substituting this value into (4.7) we have

$$\hat{\alpha}_1 = \frac{\sum\limits_{t=1}^{N-1} (x_t - \bar{x})(x_{t+1} - \bar{x})}{\sum\limits_{t=1}^{N-1} (x_t - \bar{x})^2} \tag{4.9}$$

It is interesting to note that this is exactly the same estimator that would arise if we were to treat the autoregressive equation

$$X_t - \bar{x} = \alpha_1 (x_{t-1} - \bar{x}) + Z_t$$

as an ordinary regression with $(x_{t-1} - \bar{x})$ as the 'independent' variable. In fact H. B. Mann and A. Wald showed in 1943 that, asymptotically, much of classical regression theory can be applied to autoregressive situations.

A further approximation which is often used is obtained by noting that the denominator of (4.9) is approximately

$$\sum\limits_{t=1}^{N} (x_t - \bar{x})^2$$

so that

$$\hat{\alpha}_1 \simeq c_1/c_0$$

$$= r_1$$

This approximate estimator for $\hat{\alpha}_1$ is also intuitively appealing since r_1 is an estimator for $\rho(1)$ and $\rho(1) = \alpha_1$ for a first-order AR process. A confidence interval for α_1 may be obtained from the fact that the asymptotic standard error of $\hat{\alpha}_1$ is $\sqrt{\{(1 - \alpha_1^2)/N\}}$, although the confidence interval will not be symmetric for $\hat{\alpha}_1$ away from zero. When $\alpha_1 = 0$, the standard error of $\hat{\alpha}_1$ is $1/\sqrt{N}$, and so a test for $\alpha_1 = 0$ is given by seeing if $\hat{\alpha}_1 = r_1$ lies within the range $\pm 2/\sqrt{N}$. This is equivalent to the test for $\rho(1) = 0$ already noted in Section 4.1.

For a second-order AR process, with $p = 2$, similar approximations may be made to give

$$\hat{\mu} \simeq \bar{x}$$

$$\hat{\alpha}_1 \simeq r_1(1 - r_2)/(1 - r_1^2) \tag{4.10}$$

$$\hat{\alpha}_2 \simeq (r_2 - r_1^2)/(1 - r_1^2) \tag{4.11}$$

These results are also intuitively reasonable in that if we fit a second-order model to what is really a first-order process, then as $\alpha_2 = 0$ we have $\rho(2) = \rho(1)^2 = \alpha_1^2$ and so $r_2 \simeq r_1^2$. Thus equations (4.10) and (4.11) become $\hat{\alpha}_1 \simeq r_1$ and $\hat{\alpha}_2 \simeq 0$. Jenkins and Watts (1968, p. 197) describe $\hat{\alpha}_2$ as the (sample) **partial** autocorrelation coefficient of order two which measures the excess correlation between $\{X_t\}$ and $\{X_{t+2}\}$ not accounted for by r_1.

In addition to point estimates of α_1 and α_2 it is also possible to find a confidence region in the (α_1, α_2) plane (Jenkins and Watts, 1968, p. 192).

Higher-order AR processes may also be fitted by least squares in a straightforward way. Two alternative approximate methods are commonly used. Both methods involve taking $\hat{\mu} = \bar{x}$. The first method fits the data to the model

$$X_t - \bar{x} = \alpha_1(x_{t-1} - \bar{x}) + \cdots + \alpha_p(x_{t-p} - \bar{x}) + Z_t$$

treating it as if it were an ordinary regression model. A standard multiple regression computer program may be used with appropriate modification.

The second method involves substituting the sample autocorrelation coefficients into the first p Yule-Walker equations (see Section 3.4.4) and solving for $(\hat{\alpha}_1, \ldots, \hat{\alpha}_p)$ (e.g. Pagano, 1972). In matrix form these equations are

$$R\hat{\alpha} = r \tag{4.12}$$

where

$$R = \begin{pmatrix} 1 & r_1 & r_2 & \cdots & r_{p-1} \\ r_1 & 1 & r_1 & \cdots & r_{n-2} \\ r_2 & r_1 & 1 & \cdots & r_{p-3} \\ \cdots & \cdots & \cdots & \cdots & \cdots \\ r_{p-1} & r_{p-2} & & \cdots & 1 \end{pmatrix}$$

is a $(p \times p)$ matrix,

$$\hat{\alpha}^T = (\hat{\alpha}_1, \ldots, \hat{\alpha}_p)$$

and

$$r^T = (r_1, \ldots, r_p)$$

For N reasonably large, both methods will give estimated values 'very close' to the true least squares estimates for which $\hat{\mu}$ is close to but not necessarily equal to $\bar{x}$.

4.2.2   Determining the order of an autoregressive process

It is usually difficult to assess the order of an AR process from the sample ac.f. alone. For a first-order process the theoretical ac.f. decreases exponentially and the sample function should have a similar shape. But for higher-order processes the ac.f. may be a mixture of damped exponential or sinusoidal functions and is difficult to identify. One approach is to fit AR processes of progressively higher order, to calculate the residual sum of squares for each value of p, and to plot this against p. It may then be possible to see the value of p where the curve 'flattens out' and the addition of extra parameters gives little improvement in fit.

Another aid to determining the order of an AR process is the **partial autocorrelation function** (see Box and Jenkins, 1970, p. 64) which is defined as follows. When fitting an AR(p) model, the last coefficient α_p will be denoted by π_p and measures the excess correlation at lag p which is not accounted for by an AR($p-1$) model. It is called the pth partial autocorrelation coefficient and, when plotted against p, gives the partial ac.f. The first partial autocorrelation coefficient π_1 is simply equal to $\rho(1)$, and this is equal to α_1 for an AR(1) process. It can be shown (see Exercise 4.3) that the second partial correlation coefficient is $[\rho(2)-\rho(1)^2]/[1-\rho(1)^2]$, and we note that this is zero for an AR(1) process where $\rho(2)=\rho(1)^2$.

The sample partial ac.f. is estimated by fitting AR processes of successively higher order and taking $\hat{\pi}_1 = \hat{\alpha}_1$ when an AR(1) process is fitted, taking $\hat{\pi}_2 = \hat{\alpha}_2$ when an AR(2) process is fitted, and so on. Values of $\hat{\pi}_p$ which are outside the range $\pm 2/\sqrt{N}$ are significantly different from zero at the 5% level. It can be shown that the partial ac.f. of an AR(p) process 'cuts off' at lag p so that the 'correct' order is assessed as that value of p beyond which the sample values of $\{\pi_j\}$ are not significantly different from zero. In contrast the partial ac.f. of an MA process will generally attenuate, and so the partial ac.f. has 'opposite' properties to the ac.f.

Some additional tools to help in model identification are discussed in Section 11.4.

4.3   FITTING A MOVING AVERAGE PROCESS

Suppose now that an MA process is thought to be an appropriate model for a given time series. As for an AR process, we have two problems:

(a) Finding the order of the process
(b) Estimating the parameters of the process.

We consider problem (b) first.

4.3.1   Estimating the parameters of a moving average process

Estimation problems are more difficult for an MA process than an AR process, because efficient explicit estimators cannot be found. Instead some form of numerical iteration must be performed.

Let us begin by considering the first-order MA process

$$X_t = \mu + Z_t + \beta_1 Z_{t-1} \tag{4.13}$$

where μ, β_1 are constants and Z_t denotes a purely random process. We would like to write the residual sum of squares, ΣZ_t^2, solely in terms of the observed xs and the parameters μ, β_1, as we did for the AR process, to differentiate with respect to μ and β_1, and hence to find the least squares estimates. Unfortunately the residual sum of squares is not a quadratic function of the parameters and so explicit least squares estimates cannot be found. Nor can we simply equate sample and theoretical first-order autocorrelation coefficients by

$$r_1 = \hat{\beta}_1/(1 + \hat{\beta}_1^2) \tag{4.14}$$

and choose the solution $\hat{\beta}_1$ such that $|\hat{\beta}_1| < 1$, because it can be shown that this gives rise to an inefficient estimator.

The approach suggested by Box and Jenkins (1970, Chapter 7) is as follows. Select suitable starting values for μ and β_1, such as $\mu = \bar{x}$ and β_1 given by the solution of equation (4.14) (see Table A in Box and Jenkins, 1970). Then the corresponding residual sum of squares may be calculated using (4.13) recursively in the form

$$Z_t = X_t - \mu - \beta_1 Z_{t-1} \tag{4.15}$$

With $z_0 = 0$, we have

$$z_1 = x_1 - \mu, \qquad z_2 = x_2 - \mu - \beta_1 z_1, \ldots,$$

$$z_N = x_N - \mu - \beta_1 z_{N-1}$$

Then $\sum_{t=1}^{N} z_t^2$ may be calculated.

This procedure could then be repeated for other values of μ and β_1 and the sum of squares Σz_t^2 computed for a grid of points in the (μ, β_1) plane. We may then determine by inspection the least squares estimates of μ and β_1 which minimize Σz_t^2. These least squares estimates are also maximum likelihood estimates conditional on the fixed zero value for z_0 provided that Z_t is normally distributed. The procedure can be further refined by **back forecasting** the value of z_0 (see Box and Jenkins, 1970), but this is unnecessary except when N is small or when β_1 is 'close' to plus or minus one. Nowadays the values of μ and β which minimize Σz_t^2 would normally be found by some iterative optimization procedure, such as hill-climbing, although a grid search can still sometimes be useful to see what the sum of squares surface looks like.

An alternative estimation procedure due to J. Durbin is to fit a high-order AR process to the data and use the duality between AR and MA processes (see for example Kendall, Stuart and Ord, 1983, Section 50.16). This procedure has the advantage of requiring less computation, but the widespread availability of high-speed computers has resulted in the procedure becoming obsolete.

For higher-order processes a similar type of iterative procedure to that described above may be used. For example, with a second-order MA process one would guess starting values for μ, β_1, β_2, compute the residuals recursively using

$$z_t = x_t - \mu - \beta_1 z_{t-1} - \beta_2 z_{t-2}$$

and compute Σz_t^2. Then other values of μ, β_1, β_2 could be tried, perhaps over a grid of points, until the minimum value of Σz_t^2 is found. Clearly a computer is essential for performing such a large number of arithmetic operations, and a numerically efficient optimization procedure is often used to minimize the residual sum of squares. Box and Jenkins (1970, Section 7.2) describe such a procedure, which they call 'non-linear estimation'. This description arises from the fact that the residuals are non-linear functions of the parameters (see Section 11.3), but the description may give rise to confusion.

For a completely new set of data, it may be a good idea to use the method based on evaluating the residual sum of squares at a grid of points. A visual examination of the sum of squares surface will sometimes provide useful information. In particular it is interesting to see how 'flat' the surface is; if the surface is approximately quadratic; and if the parameter estimates are approximately uncorrelated.

In addition to point estimates, an approximate confidence region for the model parameters may be found as described by Box and Jenkins (1970, p. 228) by assuming that the Z_t are normally distributed. But there is some doubt as to whether the asymptotic normality of maximum likelihood estimators will apply even for moderately large sample sizes (e.g. $N = 200$).

It should now be clear that it is much harder to estimate the parameters of an MA model than those of an AR model, as the 'errors' in an MA model are non-linear functions of the parameters and iterative methods are required to minimize the residual sum of squares. Because of this, many analysts prefer to fit an AR model to a given time series even though the resulting model may contain more parameters than the 'best' MA model. Indeed the relative simplicity of AR modelling is the main reason for its use in the stepwise autoregression forecasting technique (see Section 5.2.5) and in autoregressive spectrum estimation (see Section 11.5).

4.3.2   Determining the order of a moving average process

If an MA process is thought to be appropriate for a given set of data, the order of the process is usually evident from the sample ac.f. The theoretical ac.f. of an MA(q) process has a very simple form in that it 'cuts off' at lag q (see Section 3.4.3), and so the analyst should look for the lag beyond which the values of r_k are close to zero. The partial ac.f. is generally of little help in identifying MA models because of its attenuated form.

4.4 ESTIMATING THE PARAMETERS OF AN ARMA MODEL

Suppose now that a mixed autoregressive/moving-average (ARMA) model is thought to be appropriate for a given time series. The estimation problems for an ARMA model are similar to those for an MA model in that an iterative procedure has to be used. The residual sum of squares can be calculated at every point on a suitable grid of the parameter values, and the values which give the minimum sum of squares may then be assessed. Alternatively some sort of optimization procedure may be used.

As an example, consider the ARMA(1, 1) process whose ac.f. decreases exponentially after lag 1 (see Exercise 3.11). This model may be recognized as appropriate if the sample ac.f. has a similar form. The model is given by

$$X_t - \mu = \alpha_1(X_{t-1} - \mu) + Z_t + \beta_1 Z_{t-1}$$

Given N observations $x_1, \ldots, x_N$, we guess values for μ, α_1, β_1, set $z_0 = 0$ and $x_0 = \mu$, and then calculate the residuals recursively by

$$z_1 = x_1 - \mu$$
$$z_2 = x_2 - \mu - \alpha_1(x_1 - \mu) - \beta_1 z_1$$
$$\ldots\ldots\ldots\ldots$$
$$z_N = x_N - \mu - \alpha_1(x_{N-1} - \mu) - \beta_1 z_{N-1}$$

The residual sum of squares $\sum_{t=1}^{N} z_t^2$ may then be calculated. Then other values of μ, α_1, β_1 may be tried until the minimum residual sum of squares is found. Further details may be found in Box and Jenkins (1970).

Many variants of the above estimation procedure have been studied – see the reviews by Priestley (1981, Chapter 5) and Kendall, Stuart and Ord (1983, Chapter 50). Nowadays exact maximum likelihood estimates are often preferred, despite the extra computation involved. The conditional least squares estimates introduced above are conceptually easier to understand and can also be used as starting values for exact maximum likelihood. The Hannan-Rissanen recursive regression procedure (e.g. see Granger and Newbold, 1986) is primarily intended for model identification but can alternatively be used to provide starting values as well. The Kalman filter (see Section 10.1.4) may be used to calculate exact maximum likelihood estimates to any desired degree of approximation. We will say no more about this important, but rather advanced, topic here. Many computer packages now incorporate sound estimation routines.

4.5 ESTIMATING THE PARAMETERS OF AN ARIMA MODEL

In practice most time series are non-stationary, and the stationary models we have so far considered are not immediately appropriate. We can difference an observed time series until it is stationary, as described in Section 3.4.6. An AR, MA or ARMA model may then be fitted to the differenced series as described in Sections 4.2–4.4. The resulting model for the undifferenced series is the fitted ARIMA model, and two examples are given in Appendix D.

4.6 THE BOX-JENKINS SEASONAL (SARIMA) MODEL

In practice, many time series contain a seasonal periodic component which repeats every s observations. For example, with monthly observations, where $s = 12$, we may typically expect X_t to depend on terms such as X_{t-12}, and perhaps X_{t-24}, as well as terms such as X_{t-1}, X_{t-2}, ... Box and Jenkins (1970) have generalized the ARIMA model to deal with seasonality, and define a general multiplicative seasonal ARIMA model (abbreviated SARIMA model) as

$$\phi_p(B)\Phi_P(B^s)W_t = \theta_q(B)\Theta_Q(B^s)Z_t \qquad (4.16)$$

where B denotes the backward shift operator, ϕ_p, Φ_P, θ_q, Θ_Q are polynomials of order p, P, q, Q respectively, Z_t denotes a purely random process, and

$$W_t = \nabla^d \nabla_s^D X_t \qquad (4.17)$$

This model looks rather complicated at first sight. However, if say $P = 1$, then the term $\Phi_P(B^s)$ will be $(1 - \text{constant} \times B^s)$, which simply means that W_t will depend on W_{t-s}, since $B^s W_t = W_{t-s}$. The variables $\{W_t\}$ are formed from the original series $\{X_t\}$ not only by simple differencing (to remove trend) but also by seasonal differencing, ∇_s, to remove seasonality. For example if $d = D = 1$ and $s = 12$, then

$$W_t = \nabla\nabla_{12}X_t = \nabla_{12}X_t - \nabla_{12}X_{t-1}$$
$$= (X_t - X_{t-12}) - (X_{t-1} - X_{t-13})$$

The model in equations (4.16) and (4.17) is said to be a SARIMA model of order $(p, d, q) \times (P, D, Q)_s$. The values of d and D do not usually need to exceed one.

As an example, consider a SARIMA model of order $(1, 0, 0) \times (0, 1, 1)_{12}$, where we note $s = 12$. Then equations (4.16) and (4.17) can be written

$$(1 - \alpha B)W_t = (1 + \theta B^{12})Z_t$$

where $W_t = \nabla_{12}X_t$. Then we find

$$X_t = X_{t-12} + \alpha(X_{t-1} - X_{t-13}) + Z_t + \theta Z_{t-12}$$

so that X_t depends on X_{t-1}, X_{t-12} and X_{t-13} as well as the innovation at time $(t-12)$.

When fitting a seasonal model to data, the first task is to assess values of d

and D which reduce the series to stationarity and remove most of the seasonality. Then the values of p, P, q and Q need to be assessed by looking at the ac.f. and partial ac.f. of the differenced series and choosing a SARIMA model whose ac.f. and partial ac.f. are of similar form (see Section 5.2.4 and Example D.3). Finally, the model parameters may be estimated by some suitable iterative procedure. Full details are given by Box and Jenkins (1970, Chapter 9), but the many computer programs now available means that the average analyst need not worry too much about the practical details of estimation routines.

4.7 RESIDUAL ANALYSIS

When a model has been fitted to a time series, it is advisable to check that the model really does provide an adequate description of the data. As with most statistical models, this is usually done by looking at the **residuals**, which are defined by

$$\text{residual} = \text{observation} - \text{fitted value}$$

For a univariate time-series model, the fitted value is the one-step-ahead forecast so that the residual is the one-step-ahead forecast error. For example, with an AR(1) model (equation (3.4)) where α is estimated by least squares, the fitted value at time t is $\hat{\alpha}x_{t-1}$ so that the residual corresponding to x_t is

$$\hat{z}_t = x_t - \hat{\alpha}x_{t-1}$$

Of course if α were known exactly then the exact error $z_t = x_t - \alpha x_{t-1}$ could be calculated, but this situation rarely arises in practice.

If we have a 'good' model then we expect the residuals to be 'random' and 'close to zero', and model validation usually consists of plotting residuals in various ways. With time-series models we have the added feature that the residuals ordered in time and it is natural to treat them as a time series.

The two obvious steps are to plot the residuals as a time plot, and to calculate the correlogram of the residuals. The time plot will reveal any outliers and any obvious autocorrelation or cyclic effects. The residual correlogram will enable autocorrelation effects to be examined more closely. Let r_k denote the autocorrelation coefficient at lag k of the $\{\hat{z}_t\}$. If we have fitted the true model, then the true errors form a purely random process and, from Section 4.1, their correlogram is such that each autocorrelation coefficient is approximately normally distributed, mean 0, variance $1/N$ for reasonably large values of N. However, the correlogram of the residuals has somewhat different properties. For example for an AR(1) process with $\alpha = 0.7$, the 95% confidence limits are at $\pm 1.3/\sqrt{N}$ for r_1, $\pm 1.7/\sqrt{N}$ for r_2, and $\pm 2/\sqrt{N}$ for values of r_k at higher lags. Thus for lags greater than 2, the confidence limits **are** the same as for the correlogram of the true errors.

The analysis of residuals from ARMA processes is discussed by Box and Pierce (1970) and Box and Jenkins (1970, Chapter 8). It turns out that $1/\sqrt{N}$ supplies an **upper** bound for the standard error of the residual r_ks, so that values which lie outside the range $\pm 2/\sqrt{N}$ are significantly different from zero at the 5% level and give evidence that the wrong model has been fitted.

Instead of looking at the r_ks one at a time, Box and Jenkins (1970, Section 8.2.2) describe what they call a portmanteau lack-of-fit test which looks at the first M values of the correlogram all at once. The test statistic is

$$Q = N \sum_{k=1}^{M} r_k^2 \tag{4.18}$$

where N is the number of terms in the differenced series and M is typically chosen in the range 15 to 30. If the fitted model is appropriate, then Q should be approximately distributed as χ^2 with $(M-p-q)$ degrees of freedom, where p, q are the number of AR and MA terms respectively in the model. Unfortunately the χ^2 approximation can be rather poor for $N < 100$, and various alternative statistics have been proposed (e.g. Ljung and Box, 1978 suggest $N(N+2)\Sigma r_k^2/(N-k)$). However, the tests have rather poor power properties (e.g. Davies and Newbold, 1979) and in my experience rarely give significant results. A variety of other procedures have also been proposed for looking at residuals (e.g. Newbold, 1988, Section 4), but my own preference is usually just to 'look' at the few values of r_k, particularly at lags 1, 2 and the first seasonal lag (if any), and see if any are significantly different from zero using the crude limits of $\pm 2/\sqrt{N}$. If they are, then I would modify the model in an appropriate way by putting in extra terms to account for the significant coefficient(s). However, if only one (or two) values of r_k are just significant at lags which have no obvious physical meaning (e.g. $k=5$), then there would not be enough evidence to reject the model.

Another statistic which is used for testing residuals is the Durbin-Watson statistic (see Granger and Newbold, 1986, Section 6.2). This often appears in computer output and in my experience few people know what it means. The statistic is defined by

$$d = \sum_{t=2}^{N} (\hat{z}_t - \hat{z}_{t-1})^2 \bigg/ \sum_{t=1}^{N} \hat{z}_t^2 \tag{4.19}$$

Now since

$$\sum_{t=2}^{N} (\hat{z}_t - \hat{z}_{t-1})^2 \simeq 2 \sum_{t=1}^{N} \hat{z}_t^2 - 2 \sum_{t=2}^{N} \hat{z}_t \hat{z}_{t-1}$$

we find $d \simeq 2(1-r_1)$, where $r_1 = \Sigma \hat{z}_t \hat{z}_{t-1}/\Sigma \hat{z}_t^2$ is the first autocorrelation coefficient of the residuals (since the mean residual should be virtually zero). Thus the Durbin-Watson statistic is simply the residual r_1 in a different guise.

If the true model has been fitted, then we expect $r_1 \simeq 0$ and $d \simeq 2$, so that a 'typical' value for d is around two and not zero. Furthermore, a test on d is asymptotically equivalent to a test on the residual r_1.

The Durbin-Watson statistic was originally proposed for use with multiple regression models as applied to time-series data. Suppose we have N observations on a dependent variable y, and k explanatory variables $x_1, \ldots, x_k$, and we fit the model

$$Y_t = \beta_1 x_{1t} + \cdots + \beta_k x_{kt} + Z_t \qquad t = 1, \ldots, N$$

Having estimated the parameters $\{\beta_i\}$ by least squares, we want to see if the error terms are really independent. The residuals are therefore calculated by

$$\hat{z}_t = y_t - \hat{\beta}_1 x_{1t} - \cdots - \hat{\beta}_k x_{kt} \qquad t = 1, \ldots, N$$

The statistic d may now be calculated, and the distribution of d under the null hypothesis that the z_t are independent has been investigated. Tables of critical values are available (e.g. Kendall, Stuart and Ord, 1983) and they depend on the number of explanatory variables. Since d corresponds to the residual r_1, this test implies that we are only considering an AR(1) process as an alternative to a purely random process for z_t. Although it may be possible to modify the use of the Durbin-Watson statistic for models other than multiple regression models, it is usually better to look at the correlogram of the residuals as described earlier.

If the residual analysis indicates that the fitted model is inadequate in some way then the alternative models may need to be tried, and there are various tools for comparing the fit of several competing models (see Section 11.4). An iterative strategy for building time-series models, which is an integral part of the Box-Jenkins approach, is discussed more fully in Section 4.8 and 5.2.4.

4.8   GENERAL REMARKS ON MODEL BUILDING

How do we set about finding a suitable model for a given time series? The answer depends on a number of factors, including the properties of the series as assessed by a visual examination of the data, the number of observations available, and the way the model is to be used.

First it is important to understand the three main stages in model building, which can be described as:

(a) Model formulation (or model specification)
(b) Model estimation (or model fitting)
(c) Model checking (or model verification).

Textbooks often concentrate on estimation, but say little about formulation. This is unfortunate because computer packages now make estimation straightforward for many types of model, so that the real problem is knowing which model to fit in the first place. The use of residual analysis in model checking is receiving increasing attention, and may result in several cycles of model fitting as a model is modified and improved in response to these checks or in response to additional data. Thus model building is an iterative, interactive process (see also Chatfield, 1988b, Chapter 5).

This section concentrates on model formulation. The analyst should consult appropriate 'experts' about the given problem, ask questions to get relevant background knowledge, look at a time plot of the data to assess their more important features, and make sure that a proposed model is consistent with empirical and/or theoretical knowedge and with the objectives of the investigation.

There are many classes of time-series model to choose from. Chapter 3 introduced a general class of (univariate) models called ARIMA models, which includes AR, MA and ARMA models as special cases. This useful class of processes provides a good fit to many different types of time series and should generally be considered when more than about 50 observations are available. Another general class of models is the trend and seasonal type of model introduced in Chapter 2. Later in this book several more classes of model will be introduced, including multivariate models of various types, and structural models.

In areas such as oceanography and electrical engineering, long stationary series often occur. If a parametric model is required, an ARMA model should be considered and can be fitted as outlined earlier in the chapter. The observed correlogram and the partial ac.f. are examined, the appropriate ARMA model identified, and the model parameters estimated by least squares. However, as we shall see in Chapters 6 and 7, we may be more interested in the frequency properties of the time series, in which case an ARMA model may not be very helpful.

In many other areas, such as economics and marketing, non-stationary series often occur and in addition may be fairly short. If more than 50 observations are available, Box and Jenkins (1970) advocate the fitting of ARIMA models by differencing the observed time series until it becomes stationary and then fitting an ARMA model to the differenced series. For seasonal series, the seasonal ARIMA model may be used. However, it should be clearly recognized that when the variation of the systematic part of the time series (i.e. the trend and seasonality) is dominant, the effectiveness of the ARIMA model is mainly determined by the initial differencing operations and not by the subsequent fitting of an ARMA model to the differenced series, even though the latter operation is much more time-consuming. Thus the simple models discussed in Chapter 2 may be preferable for time series with a pronounced trend and/or large seasonal effect. Models of this type have the advantage of being simple, easy to interpret and fairly robust. In addition they can be used for short series where it is impossible to fit an ARIMA model.

EXERCISES

4.1 Derive the least squares estimates for an AR(1) process having mean μ (i.e. derive equations (4.6) and (4.7), and check the approximations in equations (4.8) and (4.9)).

4.2 Derive the least squares normal equations for an AR(p) process, taking $\hat{\mu} = \bar{x}$, and compare with the Yule-Walker equations (equation (4.12)).

4.3 Show that the (theoretical) partial autocorrelation coefficient of order 2, π_2, is given by

$$[\rho(2) - \rho(1)^2]/[1 - \rho(1)^2]$$

Compare with equation (4.11).

4.4 Find the partial ac.f. of the AR(2) process given by

$$X_t = \frac{1}{3} X_{t-1} + \frac{2}{9} X_{t-2} + Z_t$$

(see Exercise 3.6).

4.5 Suppose that the correlogram of a time series consisting of 100 observations has $r_1 = 0.31$, $r_2 = 0.37$, $r_3 = -0.05$, $r_4 = 0.06$, $r_5 = -0.21$, $r_6 = 0.11$, $r_7 = 0.08$, $r_8 = 0.05$, $r_9 = 0.12$, $r_{10} = -0.01$. Suggest an ARMA model which may be appropriate.

4.6 The first eight values of the ac.f. and partial ac.f. of 60 observations on a quarterly economic index, and of the first differences, are shown below.

Lag	1	2	3	4	5	6	7	8
X_t $\left\{ r_k \right.$	0.95	0.91	0.87	0.82	0.79	0.74	0.70	0.67
$\left. \hat{\pi}_k \right.$	0.95	0.04	−0.05	0.07	0.00	0.07	−0.04	−0.02
∇X_t $\left\{ r_k \right.$	0.02	0.08	0.12	0.05	−0.02	−0.05	−0.01	0.03
$\left. \hat{\pi}_k \right.$	0.02	0.08	0.06	0.03	−0.05	−0.06	−0.04	−0.02

Identify a model for the series. What else would you like to know about the data in order to make a better job of formulating a 'good' model?

5
Forecasting

Forecasting is the art of saying what will happen, and then explaining why it didn't!
Anonymous

5.1 INTRODUCTION

Forecasting the future values of an observed time series is an important problem in many areas, including economics, production planning, sales forecasting and stock control.

Suppose we have an observed time series $x_1, x_2, \ldots, x_N$. Then the basic problem is to estimate future values such as x_{N+k}, where the integer k is called the **lead time**. The forecast of x_{N+k} made at time N for k steps ahead will be denoted by $\hat{x}(N, k)$.

A wide variety of different forecasting procedures are available and it is important to realize that no single method is universally applicable. Rather the analyst must choose the procedure which is most appropriate for a given set of conditions. It is also worth bearing in mind that forecasting is a form of extrapolation, with all the dangers that entails. Forecasts are conditional statements about the future based on specific assumptions. Thus forecasts are not sacred and the analyst should always be prepared to modify them as necessary in the light of any external information. For long-term forecasting, it can be helpful to produce several different forecasts based on alternative sets of assumptions so that alternative 'scenarios' can be explored.

Forecasting methods may be broadly classified into three groups as follows.

(a) *Subjective*

Forecasts can be made on a subjective basis using judgement, intuition, commercial knowledge and any other relevant information. Methods range widely from bold freehand extrapolation to the Delphi technique, in which a group of forecasters try to obtain a consensus forecast with controlled feedback of other analysts' preliminary predictions. These methods will not be described here (see e.g. Armstrong, 1985; Wright and Ayton, 1987) as most statisticians will want their forecasts to be at least partly objective. However, note that some subjective judgement is often used in a more statistical approach, for example to choose an appropriate model and perhaps make adjustments to the resulting forecasts.

(b) *Univariate*

Forecasts of a given variable are based on a model fitted only to past observations of the given time series, so that $\hat{x}(N, k)$ depends only on $x_N, x_{N-1}, \ldots$. For example, forecasts of future sales of a given product would be based entirely on past sales. Methods of this type are sometimes called naive or projection methods.

(c) *Multivariate*

Forecasts of a given variable depend at least partly on values of one or more other series, called predictor or explanatory variables. For example, sales forecasts may depend on stocks and/or on economic indices. Methods of this type are sometimes called causal models.

In practice, a forecasting procedure may involve a **combination** of the above approaches. In particular, marketing forecasts are often made by combining statistical predictions with the subjective knowledge and insight of people involved in the market. A more formal type of combination is to compute a weighted average of two or more objective forecasts, as this often proves superior on average to the individual forecasts. However, an informative model does not result.

   An alternative way of classifying forecasting methods is between an **automatic** approach requiring no human intervention, and a **non-automatic** approach requiring some subjective input from the forecaster. The latter applies to subjective methods and most multivariate methods. Most univariate methods can be made fully automatic but can also be used in a non-automatic form, and there can be a surprising difference between the results.

   The choice of method depends on a variety of considerations, including:

(a) How the forecast is to be used.
(b) The type of time series and its properties, such as presence/absence of trend and/or seasonality. Some series are very regular and hence 'very predictable', but others are not. As always, a time plot of the data is very helpful.
(c) How many past observations are available.
(d) The length of the forecasting horizon. This book is mainly concerned with short-term forecasting. For example, in stock control the lead time for which forecasts are required is the time between ordering an item and its delivery.
(e) The number of series to be forecast and the cost allowed per series.
(f) The skill and experience of the analyst and the computer programs available. The analyst should select a method he feels 'happy' with, and also consider the possibility of trying more than one method.

It is particularly important to clarify the objectives (as in any statistical investigation). This means finding out how a forecast will actually be used, and whether it may even influence the future. Some forecasts are self-fulfilling. In a commercial environment, forecasting should be an integral part of the management process leading to what is sometimes called a **systems approach**.

Although point forecasts are sometimes adequate, a **prediction interval** is often helpful to indicate future uncertainty. The latter can be calculated assuming that the fitted model holds true in the future or can be calculated on an empirical basis from the fitted errors (e.g. Gardner, 1988). They tend to be too narrow in practice, mainly because the underlying model may change.

Whatever forecasting method is used, some sort of forecast monitoring scheme is often advisable, particularly with large numbers of series. A variety of tracking signals for detecting 'trouble' are described by Gardner (1983).

5.2   UNIVARIATE PROCEDURES

This section introduces the many projection methods which are now available. Further details may be found in Granger and Newbold (1986), Montgomery and Johnson (1976), Abraham and Ledolter (1983) and Gilchrist (1976).

5.2.1   Extrapolation of trend curves

For long-term forecasting it is often useful to fit a **trend curve** (or **growth curve**) to successive yearly totals and extrapolate. This problem is discussed by Harrison and Pearce (1972) and Gilchrist (1976, Chapter 9). A variety of curves may be tried including polynomial, exponential, logistic and Gompertz curves (see also Section 2.5.1). At least seven to ten years of historical data are required, and Harrison and Pearce suggest that 'one should not make forecasts for a longer period ahead than about half the number of past years for which data are available.' The method is worth considering for long-term forecasting, where it is unlikely to be worthwhile to fit a complicated model to past data.

A drawback to the use of trend curves is that there is no logical basis for choosing among the different curves except by goodness-of-fit. Unfortunately it is often the case that one can find several curves which fit a given set of data almost equally well but which, when projected forward, give widely different forecasts.

5.2.2   Exponential smoothing

This forecasting procedure, first suggested by C. C. Holt in about 1958, should only be used in its basic form for non-seasonal time series showing no systematic trend. Of course many time series which arise in practice do contain a trend or seasonal pattern, but these effects can be measured and removed to produce a stationary series. Thus it turns out that adaptations of exponential smoothing are useful for many types of time series. Gardner (1985) gives a general review.

Given a non-seasonal time series with no systematic trend, $x_1, x_2, \ldots, x_N$, it is natural to take as an estimate of x_{N+1} a weighted sum of the past observations:

$$\hat{x}(N, 1) = c_0 x_N + c_1 x_{N-1} + c_2 x_{N-2} + \cdots \tag{5.1}$$

where the $\{c_i\}$ are weights. It seems sensible to give more weight to recent observations and less weight to observations further in the past. An intuitively appealing set of weights are geometric weights, which decrease by a constant ratio. In order that the weights sum to one, we take

$$c_i = \alpha(1-\alpha)^i \qquad i = 0, 1, \ldots$$

where α is a constant such that $0 < \alpha < 1$. Then (5.1) becomes

$$\hat{x}(N, 1) = \alpha x_N + \alpha(1-\alpha)x_{N-1} + \alpha(1-\alpha)^2 x_{N-2} + \cdots \tag{5.2}$$

Strictly speaking, equation (5.2) implies an infinite number of past observations, but in practice there will only be a finite number. So equation (5.2) is customarily rewritten in the **recurrence** form as

$$\hat{x}(N, 1) = \alpha x_N + (1-\alpha)[\alpha x_{N-1} + \alpha(1-\alpha)x_{N-2} + \cdots]$$
$$= \alpha x_N + (1-\alpha)\hat{x}(N-1, 1) \tag{5.3}$$

If we set $\hat{x}(1, 1) = x_1$, then equation (5.3) can be used recursively to compute forecasts. Equation (5.3) also reduces the amount of arithmetic involved since forecasts can easily be updated using only the latest observation and the previous forecast.

The procedure defined by equation (5.3) is called exponential smoothing. The adjective 'exponential' arises from the fact that the geometric weights lie on an exponential curve, but the procedure could equally well be called geometric smoothing.

Equation (5.3) is sometimes rewritten in the **error-correction** form

$$\hat{x}(N, 1) = \alpha[x_N - \hat{x}(N-1, 1)] + \hat{x}(N-1, 1)$$
$$= \alpha e_N + \hat{x}(N-1, 1) \tag{5.4}$$

where $e_N = x_N - \hat{x}(N-1, 1)$ is the prediction error at time N.

It can be shown that exponential smoothing is optimal if the underlying model for the time series is given by

$$X_t = \mu + \alpha \sum_{j<t} Z_j + Z_t \tag{5.5}$$

This infinite moving average process is non-stationary, but the first differences

$(X_{t+1} - X_t)$ form a first-order moving average process, so that X_t is an ARIMA(0, 1, 1) process (see Exercise 5.6).

The value of the smoothing constant α depends on the properties of the given time series. Values between 0.1 and 0.3 are commonly used and produce a forecast which depends on a large number of past observations. Values close to one are used rather less often and give forecasts which depend much more on recent observations. When $\alpha = 1$, the forecast is equal to the most recent observation.

The value of α may be estimated from past data by a similar procedure to that used for estimating the parameters of a moving average process. The sum of squared prediction errors is computed for different values of α and the value is chosen which minimizes the sum of squares. With a given value of α, calculate

$$\hat{x}(1, 1) = x_1$$

$$e_2 = x_2 - \hat{x}(1, 1)$$

$$\hat{x}(2, 1) = \alpha e_2 + \hat{x}(1, 1)$$

$$e_3 = x_3 - \hat{x}(2, 1)$$

$$\cdots\cdots\cdots\cdots\cdots\cdots$$

$$e_N = x_N - \hat{x}(N-1, 1)$$

and compute $\sum_{i=2}^{N} e_i^2$. Repeat this procedure for other values of α between 0 and 1, say in steps of 0.1, and select the value which minimizes Σe_i^2. Usually the sum of squares surface is quite flat near the minimum and so the choice of α is not critical.

5.2.3   The Holt-Winters forecasting procedure

Exponential smoothing may readily be generalized to deal with time series containing trend and seasonal variation. The resulting procedure is usually referred to as the Holt-Winters procedure in honour of P. R. Winters' pioneering work of 1960. Trend and seasonal terms are introduced which are also updated by exponential smoothing.

Suppose the observations are monthly, and let L_t, T_t, I_t denote the local level, trend and seasonal index, respectively, at time t. Thus T_t is the expected increase or decrease per month in the current level. Let α, γ, δ denote the three smoothing parameters for updating the level, trend and seasonal index respectively. The smoothing parameters are usually chosen in the range (0, 1). Then, when a new observation x_t becomes available, the values of L_t, T_t and I_t are all updated. If the seasonal variation is multiplicative, then the (recurrence form) updating equations are

$$L_t = \alpha(x_t/I_{t-12}) + (1-\alpha)(L_{t-1} + T_{t-1})$$

$$T_t = \gamma(L_t - L_{t-1}) + (1 - \gamma)T_{t-1}$$
$$I_t = \delta(x_t/L_t) + (1 - \delta)I_{t-12}$$

and the forecasts from time t are then

$$\hat{x}(t, k) = (L_t + kT_t)I_{t-12+k}$$

for $k = 1, 2, \ldots, 12$. There are analogous formulae for the additive seasonal case. Unfortunately the literature is confused by many different notations and by the fact that the updating equations may be presented in the equivalent error-correction form such as

$$T_t = T_{t-1} + \alpha\gamma e_t/I_{t-12}$$

where it looks as though $\alpha\gamma$ is a smoothing parameter.

A graph of the data should be examined to see if an additive or a multiplicative seasonal effect is the more appropriate. If the seasonal period does not cover 12 observations, then the updating equations need to be modified in an obvious way.

In order to implement the method, the user must

(a) Provide starting values for L_t, T_t and I_t at the beginning of the series
(b) Estimate values for α, γ, δ by minimizing Σe_t^2 over a suitable fitting period for which historical data are available
(c) Decide whether or not to normalize the seasonal indices at regular intervals (see Section 2.6)
(d) Choose between an automatic or non-automatic approach.

Full details on these and other practical questions are given by Gardner (1985) and Chatfield and Yar (1988). The method is straight forward and is widely used in practice.

5.2.4   The Box-Jenkins procedure

This section gives a brief outline of the forecasting procedure based on autoregressive integrated moving average (ARIMA) models which is usually known as the Box-Jenkins approach. The beginner may find it easier to read books such as Vandaele (1983), Granger and Newbold (1986) or Jenkins (1979) rather than the original book by Box and Jenkins (1970), although the latter is still an essential reference source.

The main stages in setting up a Box-Jenkins forecasting model are as follows.

(a) *Model identification*

Examine the data to see which member of the class of ARIMA processes appears to be most appropriate.

(b) *Estimation*

Estimate the parameters of the chosen model as described in Chapter 4.

(c) *Diagnostic checking*

Examine the residuals from the fitted model to see if it is adequate.

(d) *Consider alternative models if necessary*

If the first model appears to be inadequate for some reason, then other ARIMA models may be tried until a satisfactory model is found.

Now AR, MA and ARMA models have been around for many years and are associated in particular with G. U. Yule and H. O. Wold. The major contribution of Box and Jenkins has been to provide a general **strategy** for time-series forecasting, in which the different stages of model building as listed above are all given due prominence. In addition, they showed how the use of differencing can extend the use of ARMA models to deal with non-stationary series, and also how to incorporate seasonal terms into seasonal ARIMA (or SARIMA) models. Thus ARIMA models are often referred to as Box-Jenkins models. When a satisfactory ARIMA model has been found, it is relatively straightforward to calculate forecasts as conditional expectations.

The first step in the Box-Jenkins procedure is to difference the data until they are stationary. This is achieved by examining the correlograms of various differenced series until one is found which comes down to zero 'fairly quickly' and from which any seasonal cyclic effect has been largely removed, although there may still be 'spikes' at lags, s, $2s$, and so on, where s is the number of observations per year. For non-seasonal data, first-order differencing is usually sufficient. For seasonal data of period 12, the operator $\nabla\nabla_{12}$ is often used if the seasonal effect is additive, while the operator ∇_{12}^2 may be used if the seasonal effect is multiplicative. Sometimes the operator ∇_{12} by itself will be sufficient. Over-differencing should be avoided. For quarterly data the operator ∇_4 may be used, and so on.

The differenced series will be denoted by $\{w_t;\ t=1,\ldots,N-c\}$, where c terms are 'lost' by differencing. For example, if the operator $\nabla\nabla_{12}$ is used, then $c=13$.

If the data are non-seasonal, an ARMA model can now be fitted to $\{w_t\}$ as described in Chapter 4. If the data are seasonal, then the SARIMA model defined in equation (4.16) may be fitted as follows. 'Reasonable' values of p, P, q, Q are selected by examining the correlogram and the partial ac.f. of the differenced series $\{w_t\}$. Values of p and q are selected as outlined in Chapter 4 by examining the first few values of r_k. Values of P and Q are selected primarily by examining the values of r_k at $k=12, 24\ldots$ (where the seasonal period is 12). If for example r_{12} is 'large' but r_{24} is 'small', this suggests one seasonal moving average term, so we would take $P=0$, $Q=1$, as this SARIMA model has an ac.f. of similar form. Box and Jenkins (1970) list the acv.f.s of various SARIMA models.

Having tentatively identified what appears to be a reasonable SARIMA model, least squares estimates of the model parameters may be obtained by minimizing the residual sum of squares in a similar way to that proposed for ordinary ARMA models. In the case of seasonal series, it is advisable to estimate initial values of a_t and w_t by backforecasting (or backcasting) rather than set them equal to zero. This procedure is described by Box and Jenkins (1970, Section 9.2.4). In fact if the model contains a seasonal moving average parameter which is close to one, several cycles of forward and backward iteration may be needed. Nowadays several alternative estimation procedures are available, based for example on the exact likelihood function, on conditional or unconditional least squares, or on a Kalman filter approach (see references in Section 4.4).

For both seasonal and non-seasonal data, the adequacy of the fitted model should be checked by what Box and Jenkins call 'diagnostic checking'. This essentially consists of examining the residuals from the fitted model to see if there is any evidence of non-randomness. The correlogram of the residuals is calculated and we can then see how many coefficients are significantly different from zero and whether any further terms are indicated for the ARIMA model. If the fitted model appears to be inadequate, then alternative ARIMA models may be tried until a satisfactory one is found. Section 11.4 describes some additional model identification tools to help in choosing an appropriate model.

When a satisfactory model is found, forecasts may readily be computed. Given data up to time N, these forecasts will involve the observations and the fitted residuals (i.e. the one-step-ahead forecast errors) up to and including time N. The minimum mean square error forecast of X_{N+k} at time N is the conditional expectation of X_{N+k} at time N, namely $\hat{x}(N, k) = E(X_{N+k}|X_N, X_{N-1}, \ldots)$. In evaluating this conditional expectation, we use the fact that the 'best' forecast of all future Zs is simply zero (or more formally that the conditional expectation of Z_{N+k}, given data up to time N, is zero for all $k>0$). Box and Jenkins (1970) describe three general approaches to computing forecasts.

(a) *Using the difference equation form*

Forecasts are usually computed most easily directly from the model equation which Box and Jenkins (1970) call the difference equation form. Assuming that the model equation is known exactly, then $\hat{x}(N, k)$ is obtained from the model equation by replacing (1) future values of Z by zero, (2) future values of X by their conditional expectation, and (3) past values of X and Z by their observed values.

For example, consider the SARIMA$(1, 0, 0)(0, 1, 1)_{12}$ model used as an example in Section 4.6, where

$$X_t = X_{t-12} + \alpha(X_{t-1} - X_{t-13}) + Z_t + \theta Z_{t-12}$$

Then we find

$$\hat{x}(N, 1) = x_{N-11} + \alpha(x_N - x_{N-12}) + \theta z_{N-11}$$

$$\hat{x}(N, 2) = x_{N-10} + \alpha[\hat{x}(N, 1) - x_{N-11}] + \theta z_{N-10}$$

Forecasts further into the future can be calculated recursively in an obvious way. It is also possible to find ways of updating the forecasts as new observations become available. For example, when x_{N+1} becomes known we have

$$\hat{x}(N+1, 1) = x_{N-10} + \alpha(x_{N+1} - x_{N-11}) + \theta z_{N-10}$$

$$= \hat{x}(N, 2) + \alpha[x_{N+1} - \hat{x}(N, 1)]$$

$$= \hat{x}(N, 2) + \alpha z_{N+1}$$

(b) *Using the ψ weights*

The ψ weights defined in equation (3.6b) can also be used to compute forecasts and are particularly helpful in calculating forecast error variances. Since

$$X_{N+k} = Z_{N+k} + \psi_1 Z_{N+k-1} + \cdots$$

it is clear that $\hat{x}(N, k)$ is equal to $\sum_{j=0}^{\infty} \psi_{k+j} z_{N-j}$ (i.e. no future zs are included). Thus the k-steps-ahead forecast error is $(Z_{N+k} + \psi_1 Z_{N+k-1} + \cdots + \psi_{k-1} Z_{N+1})$. Hence the variance of the k-steps-ahead error is $(1 + \psi_1^2 + \cdots + \psi_{k-1}^2)\sigma_z^2$.

(c) *Using the π weights*

The π weights defined in equation (3.6c) can also be used. Since

$$X_{N+k} = \pi_1 X_{N+k-1} + \cdots + \pi_k X_N + \cdots + Z_{N+k}$$

it is intuitively clear that $\hat{x}(N, k)$ is given by

$$\hat{x}(N, k) = \pi_1 \hat{x}(N, k-1) + \pi_2 \hat{x}(N, k-2) + \cdots + \pi_k X_N + \pi_{k+1} X_{N-1} + \cdots$$

These forecasts can be computed recursively, replacing future values of X with predicted values.

In practice the model is not known exactly, and we have to estimate the model parameters (and hence the ψs and πs if needed); we also have to estimate the past values of Z by the observed residuals or one-step-ahead errors. Thus for the SARIMA$(1, 0, 0)(0, 1, 1)_{12}$ model given above, we would have, for example, that

$$\hat{x}(N, 1) = x_{N-11} + \hat{\alpha}(x_N - x_{N-12}) + \hat{\theta}\hat{z}_{N-11}$$

Except for short series, this generally makes little difference to forecast error variances.

Although some packages have been written to provide automatic ARIMA modelling (with dubious results), the method is primarily intended for a non-automatic approach where the analyst uses his subjective judgement to select an appropriate model from the large family of ARIMA models according to the properties of the individual series being analysed. Thus, although the procedure is more versatile than many competitors, it is also more complicated and considerable experience is required to identify an appropriate ARIMA model. Unfortunately, the analyst may find several different models which fit the data equally well but give rather different forecasts, while sometimes it is difficult to find any sensible model. The inexperienced analyst will sometimes choose a 'silly' model. Another drawback is that the method requires at least 50 observations to have a chance of success.

My own view (see also Section 5.4) is that the method should not be used by analysts with limited statistical experience or for series where the variation is dominated by trend and seasonal variation (see Example 5.2 and Example D.3). However, it can work well for series showing short-term correlation (see Example D.2). It can also be combined with seasonal adjustment methods as in the X-11 ARIMA method (e.g. Huot, Chiu and Higginson, 1986) and generalized to the multivariate case (see Section 11.9).

5.2.5   Stepwise autoregression

Granger and Newbold (1986, Section 5.4) describe a procedure called stepwise autoregression, which can be regarded as a subset of the Box-Jenkins procedure and which has the advantage of being fully automatic. The method relies on the fact that AR models are much easier to fit than MA or ARMA models even though an AR model may require extra parameters to get as good a representation of the data.

First, differences of the data are taken to allow for non-stationarity in the mean. Then a maximum possible lag, say p, is chosen. The best autoregressive model with just one lagged variable is then found:

$$W_t = \mu + \alpha_k^{(1)} W_{t-k} + e_t^{(1)} \qquad 1 \leq k \leq p$$

where $W_t = X_t - X_{t-1}$, and $\alpha_k^{(1)}$ is the autoregression coefficient at lag k when fitting one lagged variable only. Then the best autoregressive model with 2, 3, ... lagged variables is found. The procedure is terminated when the reduction in the sum of squared residuals at the jth stage is less than some pre-assigned quantity. Thus an integrated autoregressive model is fitted which is a special case of the Box-Jenkins ARIMA class. Granger and Newbold suggest choosing $p = 13$ for quarterly data and $p = 25$ for monthly data.

5.2.6 Other methods

Several other forecasting procedures have been proposed. Brown (1963) has suggested a technique called **general exponential smoothing** which consists of fitting polynomial, sinusoidal or exponential functions to the data and finding appropriate updating formulae. One special case of this is double exponential smoothing which is applicable to series containing a linear trend. Note that Brown suggests fitting by **discounted** least squares, in which more weight is given to recent observations.

Harrison (1965) has proposed a modification of seasonal exponential smoothing which consists essentially of performing a Fourier analysis of the seasonal factors and replacing them by smoothed factors. Harrison and Stevens (1976) have proposed a technique called **Bayesian forecasting**, which depends on a model called 'the dynamic linear model' whose parameters are updated by a technique called Kalman filtering (see Chapter 10).

Another proposed technique called **adaptive filtering** (Makridakis and Wheelwright, 1977) appears to be technically unsound (Chatfield, 1978b).

Various other projection methods can also be found in the literature, including structural modelling (see Chapter 10) and Parzen's ARARMA approach (Parzen, 1982). A review of the many different methods is given by Chatfield (1988a).

5.3 MULTIVARIATE PROCEDURES

This section provides a brief introduction to multivariate forecasting procedures. Interest in such methods is growing, partly because of improvements in computing power.

5.3.1 Multiple regression

This approach uses the multiple linear regression model, where the variable of interest (say y) is linearly related to one or more other variables (say $x_1, \ldots, x_p$) which are called explanatory variables. In building a regression model, it is helpful to distinguish between explanatory variables which can or cannot be controlled, and predetermined variables such as time itself. Regression on time alone would normally be regarded as a univariate procedure. Lagged values of the response and explanatory variables may be included, but the inclusion of autoregressive terms (past values of y) changes the character of the model. Note that economists sometimes describe a regression model as an econometric model.

A description of multiple regression can be found in many statistics textbooks (e.g. Anderson, 1971) and will not be repeated here. With the general availabilty of multiple regression computer programs it is computationally easy (perhaps too easy!) to fit a multiple regression model and use it for planning or forecasting.

Multiple regression models are widely used and sometimes work well, particularly in a marketing context. But there are several dangers in the method which need to be appreciated. First, the ready availability of computer programs has resulted in a tendency to put more and more explanatory variables into the model, with dubious results. The resulting model may indeed appear to give a good fit to the available data. For example, by including as many as 20 explanatory variables it is possible to achieve a multiple correlation coefficient R^2 as high as 0.995. However, this good fit may be spurious and does not necessarily mean that the model will give good forecasts (Granger and Newbold, 1974). A more sensible number of explanatory variables is a maximum of six or seven, and it is advisable to fit the model to part of the available data and then check the model by forecasting the remainder of the data. When doing this it is important to distinguish between *ex ante* forecasts, which replace future values of the explanatory variables by forecasts (and so are true forecasts), and *ex post* forecasts, which use the true values of explanatory variables. The latter can look misleadingly good.

Explanatory variables were often called independent variables in the past, but in marketing applications the so-called 'independent' variables are usually not independent at all, and if some of them are highly correlated there may be singularity problems. It is therefore advisable to look at the correlation matrix of the 'independent' variables **before** carrying out a multiple regression so that, if necessary, some variables can be excluded. It is unnecessary for the explanatory variables to be completely independent, but large correlations should be avoided.

Another difficulty arising in multiple regression is that some crucial explanatory variables may have been held more or less constant in the past, and it is then impossible to assess their effect and include them in the model in a quantitative way. For example, a company may be considering increasing its advertising expenditure and would like to construct a model which would predict the effect on sales. But if advertising has been held relatively constant in the past, then it will be impossible to estimate the effect of advertising; yet a model which excludes advertising may be useless if advertising expenditure is changed.

But perhaps the most important problem in multiple regression forecasting concerns the structure of the error terms. It is often assumed that these are an independent white noise sequence, but such an assumption is sometimes not appropriate (Box and Newbold, 1971). Having fitted a multiple regression model, one should check the residuals for autocorrelation as described in Section 4.7. If the residuals are autocorrelated, one can try fitting a multiple regression model with autocorrelated errors by a method, due to D. Cochrane and G. H. Orcutt, which is described by Kendall, Stuart and Ord (1983, Section 51.2).

In summary, I am inclined to agree with Brown (1963, p. 77) that the use of multiple regression models can be very dangerous except in certain special cases where one has a definite reason why one series should be related to another. An example of such a situation, involving just one explanatory variable, is given in Example 5.3. Another example is given by Bhattacharyya (1974).

5.3.2 Econometric models

Econometric models (e.g. Harvey, 1981b) often assume that an economic system can be described, not by a single equation, but by a set of simultaneous equations. For example, not only do wage rates depend on prices but also prices depend on wage rates. Economists distinguish between exogenous variables, which affect the system but are not themselves affected, and endogenous variables, which interact with each other. The simultaneous equation system involving k dependent (endogenous) variables, $\{Y_i\}$, and g predetermined (exogenous) variables, $\{X_i\}$, may be written

$$Y_i = f_i(Y_1, \ldots, Y_{i-1}, Y_{i+1}, \ldots, Y_k, X_1, \ldots, X_g) + \text{error}$$
$$i = 1, 2, \ldots, k$$

Some of the exogenous variables may be lagged values of the Y_i. These equations, often called the structural form of the system, can be solved to give what is called the reduced form of the system, namely

$$Y_i = F_i(X_1, \ldots, X_g) + \text{error} \qquad i = 1, 2, \ldots, k$$

The principles and problems involved in constructing econometric models are too broad to be discussed in detail here (e.g. see Granger and Newbold, 1986, Section 6.3). A key issue is the extent to which the form of the model should be based on judgement, on economic theory and/or on empirical data. While some econometricians have been scornful of univariate time-series models which do not 'explain' what is going on, statisticians have been generally sceptical of traditional econometric model building in which the structure of the model is determined by economic theory and little attention is paid to the 'error' structure. However, the uncontrolled nature of much economic data makes it difficult to construct econometric models solely on an empirical basis. Fortunately, mutual understanding has improved in recent years as developments in multivariate time-series modelling have brought statisticians and econometricians closer together to the benefit of both. It is now widely recognized that econometric model building should be an iterative process involving both theory **and** data.

5.3.3 Other multivariate models

There are many other types of multivariate model which may be used to produce forecasts. The multivariate generalization of ARIMA models is considered in Section 11.9. One special case is the class of transfer function models (see Section 9.4.2) which concentrates on describing the relationship between one 'output' variable and one or more 'input' or explanatory variables. It is helpful to understand the interrelationships between all these classes of multivariate model (e.g. see Granger and Newbold, 1986, Chapters 6–8; Priestley, 1981, Chapter 9).

Of course, more specialized multivariate models may occasionally be required. For example, forecasts of births must take account of the number and age of women of child-bearing age. Common sense and background knowledge of the problem should indicate what is required.

5.4 A COMPARATIVE REVIEW OF FORECASTING PROCEDURES

We noted in Section 5.1 that there is no 'best' forecasting procedure, but rather that the choice of method depends on a variety of factors such as the objective in producing forecasts, the degree of accuracy required, and the properties of the given time series. This section attempts a brief review of recent research but makes no attempt to be exhaustive. The extensive annotated list of references given by Armstrong (1985) indicates the growing research activity.

Many forecasts are used for planning purposes, while others act as a 'norm' against which the effect of changes in strategy may be assessed. Sometimes more than one forecast is required to assess the effects of different assumptions or strategies.

Univariate forecasts are particularly suitable when there are large numbers of series to be forecast (e.g. in stock control) so that a relatively simple method has to be used. They are also suitable when the analyst's skill is limited, when a 'norm' is required, or when they are otherwise judged appropriate for the client's needs and level of understanding. Multivariate models are appropriate to assess the effects of explanatory variables, to understand the economy, and to evaluate alternative economic policy proposals by constructing 'what-if' forecasts.

5.4.1 Forecasting competitions

In order to clarify the choice between different univariate methods, there have been several 'competitions' to compare the forecasting accuracy of different methods on a given set of time series. The four major competitions are described by Reid (1975), Newbold and Granger (1974), Makridakis and

Hibon (1979) and Makridakis *et al.* (1984). The last study, commonly known as the M-competition, was designed to be more wide ranging than earlier studies, and compared 24 methods on 1001 series. Given different analysts and data sets, it is perhaps not too surprising that the results from different competitions have not always been consistent. For example, the two earlier studies found that Box-Jenkins tended to give more accurate forecasts than other univariate methods, but this was not the case in the later studies. A detailed assessment of the strengths and weaknesses of forecasting competitions is given by Chatfield (1988a). It is essential that results be replicable and that appropriate criteria are used. Moreover, accuracy is only one aspect of forecasting, and practitioners think that cost, ease of use and ease of interpretation are of almost equal importance. Furthermore, competitions mainly analyse large numbers of series in a completely automatic way. Thus although they tell us something, competitions only tell part of the story and are mainly concerned with comparing automatic forecasts.

If an automatic approach is desirable or unavoidable, perhaps because a large number of series is involved, then my interpretation of the competition results is as follows. While there could be significant gains in being selective, most users will want to apply the same method to all series for obvious practical reasons. Some methods should be discarded, but there are several automatic methods for which average differences in accuracy are small. Thus the choice between them may depend on other practical considerations such as availability of computer programs. The methods include Holt's exponential smoothing, Holt-Winters and Bayesian forecasting. My particular favourite, partly on grounds of familiarity, is the Holt-Winters method, which can be recommended as a generally reliable, easy to understand, all-purpose automatic method.

5.4.2 Choosing a non-automatic method

Suppose instead that a non-automatic approach is indicated because the number of series is small and/or because external information is available which cannot be ignored. Then sensible forecasters will use their skill and knowledge to interact with their clients, incorporate background knowledge, plot the data and generally use all relevant information to build a model and compute forecasts. The choice then lies between some form of multivariate method and a non-automatic univariate procedure. Here forecasting competitions are of limited value and it is easy to cite case studies where subjective adjustment of automatic forecasts leads to improvements (e.g. Chatfield, 1978a). Moreover, the average differences in accuracy for different methods are relatively small compared with the large differences in accuracy which can arise when the methods are applied to individual series. The rewards in being selective indicate that the distinction between an automatic and a non-automatic **approach** may be more fundamental than the differences between different forecasting **methods**.

It is also easy to cite case studies (e.g. Jenkins and McLeod, 1982) where statistician and client collaborate to develop a successful multivariate model. However, it is difficult to make general statements about the relative accuracy of different multivariate methods. Many people expect multivariate forecasts to be at least as good as univariate forecasts, but this is not true either in theory or in practice, partly because the computation of multivariate forecasts may require the prior computation of forecasts of exogenous variables, and the latter may not be good enough (Ashley, 1988). Chatfield (1988a) reviews the empirical evidence. Regression models do rather better on average than univariate methods, though not by any means in every case (Fildes, 1985). Econometric simultaneous equation models have a patchy record and it is easy to cite cases where univariate forecasts are more accurate (e.g. Naylor, Seaks and Wichern, 1972; Makridakis and Hibon, 1979, Section 2). There have been some encouraging case studies using transfer function models (e.g. Jenkins, 1979; Jenkins and McLeod, 1982) but such models rely on the absence of feedback which may not apply to much economic data. Multivariate ARIMA models also have a mixed record and are perhaps more useful for understanding relationships than for forecasting. Of course multivariate models can usually be made to give a better **fit** to given data than univariate models, but this superiority does not necessarily translate into better forecasts, perhaps because multivariate models are more sensitive to changes in structure.

It has to be realized that the nature of economic time-series data is such as to make it difficult to fit reliable multivariate time-series models. Most economic variables are simply observed, rather than controlled, and there are usually high autocorrelations within each series. In addition there may be high correlations beween series, not necessarily because of a real relationship but simply because of mutual correlations with time (Pierce, 1977). Feedback between 'output' and 'input' variables is another problem. There are special difficulties in fitting regression models to time-series data anyway, as already noted in Section 5.3.1, and an apparent good fit may be spurious. Simultaneous equation and multivariate ARIMA models are even more difficult to construct, and their use seems likely to be limited to the analyst who is as interested in the modelling process as in forecasting. Thus although the much greater effort required to construct multivariate models will sometimes prove fruitful, there are many situations where a univariate method will be preferred.

With a non-automatic univariate approach, the main choice is between the Box-Jenkins approach and the non-automatic use of a simple method, such as Holt-Winters, which is perhaps more often used in automatic mode. The Box-Jenkins approach has been one of the most influential developments in time-series analysis. However, the accuracy of the resulting forecasts has been rather mixed in practice, particularly when one realizes that forecasting competitions are biased in favour of Box-Jenkins by implementing other methods in a completely automatic way. The advantage of being able to choose from the broad class of ARIMA models is clear, but, as noted in Section 5.2.4, there are also dangers in that considerable experience is needed to interpret correlograms and other indicators. Moreover, when the variation in a series is dominated by trend and seasonality, the effectiveness of the fitted ARIMA model is mainly determined by the differencing procedure rather than by the identification of the autocorrelation structure of the differenced (stationary) series, which is what is emphasized in the Box-Jenkins approach. Nevertheless, some writers have suggested that all exponential smoothing models should be regarded as special cases of Box-Jenkins, the implication being that one might as well use Box-Jenkins. However, this view is now discredited (Chatfield and Yar, 1988) because exponential smoothing methods are actually applied in a completely different way to Box-Jenkins.

In some situations, a large expenditure of time and effort can be justified and then Box-Jenkins is worth considering. However, for routine sales forecasting, simple methods are more likely to be understood by managers and workers who have to utilize or implement the results. Thus I suggest that Box-Jenkins is only worth considering when the following conditions are satisfied: (1) the analyst is competent to implement it; (2) the objectives justify the complexity; and (3) the variation in the series is **not** dominated by trend and seasonality.

5.4.3   A strategy for non-automatic univariate forecasting

If circumstances suggest a non-automatic univariate approach, then I suggest that the following steps will generally provide a sensible strategy.

(a) Get appropriate background information and carefully define the objectives.
(b) Plot the data and look for trend, seasonal variation, outliers, and changes in structure such as slow changes in variance or sudden discontinuities.
(c) 'Clean' the data if necessary, for example by adjusting any suspect observations, preferably after taking account of external information. Consider the possibility of transforming the data.
(d) Decide if the seasonal variation is (i) non-existent, (ii) multiplicative, (iii) additive or (iv) something else.
(e) Decide if the trend is (i) non-existent, (ii) global linear, (iii) local linear or (iv) non-linear.
(f) Fit an appropriate model where possible. It is helpful to distinguish four types of series:
(i) Discontinuities present. Figure 5.1(a) shows a series containing a

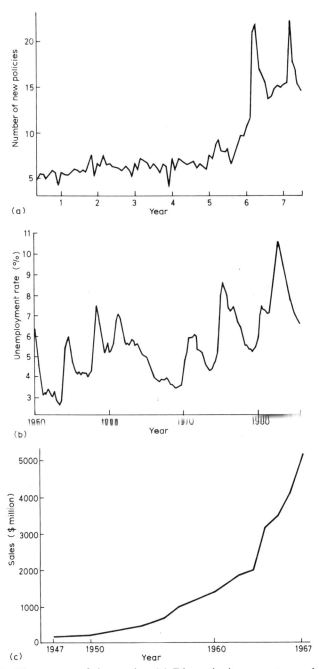

Figure 5.1 Three types of time series. (a) Discontinuity present: numbers of new insurance policies issued by a particular life office (monthly in hundreds). (b) Short-term autocorrelation present: unemployment rate in USA (quarterly). (c) Exponential growth present: world-wide sales of IBM (yearly).

major discontinuity where it is generally unwise to produce any univariate forecasts. There is further discussion of this series in Section 11.2.

(ii) Trend and seasonality present. Figures 1.3, 5.3 and D.2 show series whose variation is dominated by trend and seasonality. Here the Holt-Winters exponential smoothing method is a suitable candidate. The correct seasonal form must be chosen and the smoothing parameters can be estimated by optimizing one-step-ahead forecasts over the period of fit. Full details are given by Chatfield and Yar (1988).

(iii) Short-term correlation present. Figure 5.1(b) shows a non-seasonal series whose variation is dominated by short-term correlation. Many economic indicator series are of this form and it is essential to try to understand the autocorrelation structure. Thus the Box-Jenkins approach is recommended here. (See Example D.2.)

(iv) Exponential growth present. Figure 5.1(c) shows a series dominated by a steadily increasing trend. Series of this type are difficult to handle because exponential forecasts are inherently unstable. No one really believes that economic growth or population size can continue to increase exponentially indefinitely. Two alternative strategies are to fit a model which explicitly includes exponential (or perhaps quadratic) growth terms, or (my preference) to fit a model to the logarithms of the data (or some other suitable transformation). There is some evidence to suggest that damping the trend will improve accuracy (Gardner and McKenzie, 1985).

(g) Check the adequacy of the fitted model. In particular, study the one-step-ahead forecast errors over the period of fit to see if they have any undesirable properties such as high autocorrelation. Modify the model if necessary.

(h) Compute forecasts. Decide if the forecasts need to be adjusted subjectively because of anticipated changes in other variables, or because of any other reason.

5.4.4 Summary

It is difficult to summarize the many empirical findings (e.g. see Makridakis, 1986, especially Exhibit 1), but I make the following general observations and recommendations:

(a) Fitting the 'best' model to historical data does not necessarily minimize post-sample forecast errors. In particular, complex models often give forecasts which are no better than simple models. The more frequent and the greater number of forecasts required, the more desirable it is to use a simple approach.

(b) Combinations of forecasts from different methods are generally better than forecasts from individual methods.

(c) The higher the level of aggregation of a series, the better is the forecast accuracy.
(d) Prediction intervals, calculated on the assumption that the model fitted to past data will also be true in the future, are generally too narrow.
(e) If an automatic univariate method is required, then the Holt-Winters method is a suitable candidate, but there are several close competitors.
(f) When a non-automatic approach is appropriate, there is a wide choice from judgemental and multivariate methods through to (univariate) Box-Jenkins and the 'thoughtful' use of univariate methods which are often regarded as automatic. A strategy for non-automatic univariate forecasting has been proposed which may incorporate Box-Jenkins, Holt-Winters or some form of growth curve model. Whatever approach is used, the analyst should be prepared to improvise and modify 'objective' forecasts using subjective judgement.

5.5   SOME EXAMPLES

In this section we discuss three sets of data, to illustrate some of the problems which arise in real forecasting situations.

Example 5.1 Figure 5.2 shows the (coded) sales of a certain company in successive quarters over 6 years. Suppose that a univariate forecast is required

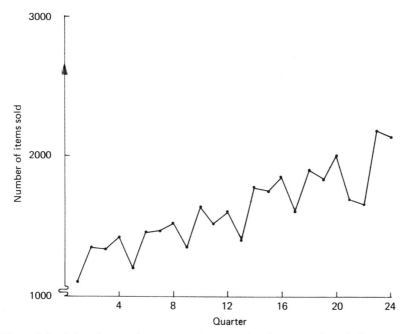

Figure 5.2   Sales of a certain company in successive three-month periods.

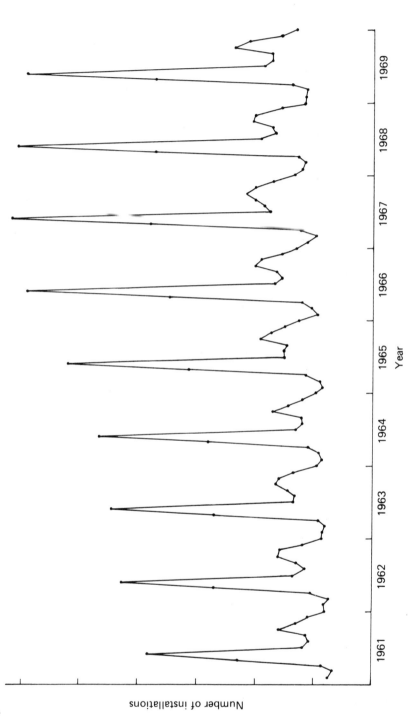

Figure 5.3 Tomasek's telephone data: numbers of newly installed telephones in successive months. (Tomasek does not provide a vertical scale.)

for the next four quarters. What method is appropriate? This series demonstrates the importance of plotting a time series and making a visual examination before deciding on the appropriate forecasting procedure. It is evident from Figure 5.2 that there is an increasing trend and a pronounced seasonal effect with observations 1, 5, 9, 13, . . . relatively low. The series is too short to use the Box-Jenkins method. Instead a suitable forecasting procedure might appear to be that of Holt-Winters. But closer examination of Figure 5.2 reveals that the observation in quarter 22 is unusually low while the following observation seems somewhat high. If we were to apply Holt-Winters with no modification, these unusual observations would have a marked effect on the forecasts. We must therefore find out if they indicate a permanent change in the seasonal pattern, in which case earlier observations will have little relevance for forecasting purposes, or if they were caused by some unusual phenomenon such as a strike, in which case some data adjustment may be advisable. Asking questions to get background information is most important.

Example 5.2   Figure 5.3 shows some telephone data analysed by Tomasek (1972) using the Box-Jenkins method. He developed the model

$$(1-0.84B)(1-B^{12})(X_t-132)=(1-0.60B)(1+0.37B^{12})Z_t$$

which, when fitted to all the data, explained 99.4% of the total variation about the mean (i.e. the total corrected sum of squares, $\Sigma(x_t-\bar{x})^2$). On the basis of this good fit, Tomasek recommended the use of the Box-Jenkins method for forecasting.

However, it is not at all clear that this is a sensible recommendation. Looking at Figure 5.3, we see that the series has an unusually high regular seasonal pattern. In fact 97% of the variation about the mean is explained by a linear trend and constant seasonal pattern. As we remarked in Section 4.8, when the variation due to trend and seasonality is dominant, the effectiveness of the ARIMA model is mainly determined by the initial differencing operations and not by the time-consuming ARMA model fitting to the differenced series (Akaike, 1973). For such regular data, nearly any forecasting method will give good results. For example, the Holt-Winters method explains 98.9% of the variation, and it is rather doubtful if the extra expense of the Box-Jenkins method can be justified by increasing the explained variation from 98.9% to 99.4%.

Example 5.3   Figure 5.4 shows quarterly sales data for company C over 12 successive years. Although there is some evidence of a seasonal pattern, it is not particularly regular. D. L. Prothero tried two univariate procedures on these data, namely Holt-Winters and Box-Jenkins. The Box-Jenkins model fitted was

$$\nabla\nabla_4 X_t = (1-0.2B)(1-0.8B^4)Z_t$$

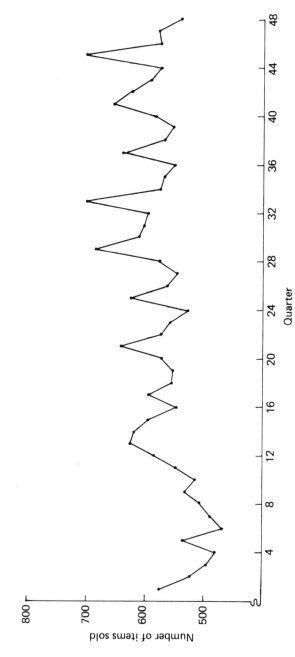

Figure 5.4 Sales of company C in successive three-month periods.

The mean absolute forecast errors up to four quarters ahead were calculated as follows:

No. of quarters ahead	1	2	3	4
Holt-Winters	25.4	28.6	31.7	37.3
Box-Jenkins	24.5	29.9	34.2	41.4

Thus although Box-Jenkins is 3% better one step ahead, it is up to 10% worse four steps ahead.

This result is typical in the sense that Box-Jenkins tends to do less well as the lead time increases. But the result is not typical in that, for series as irregular as that shown in Figure 5.4, Box-Jenkins will sometimes do considerably better than other methods. Thus, if sufficient money and expertise are available, the Box-Jenkins method is worth a try for data of this type.

These data also illustrate the possible advantages of multivariate forecasting. It was found that if detrended, deseasonalized sales are linearly regressed on detrended, deseasonalized stocks two quarters before, the mean absolute forecast error one step ahead was 19.0, which is considerably better than either of the two univariate procedures. In other words stocks are a leading indicator for sales. This illustrates the general point that if one wants to put in a lot of effort to get a good forecast, it may well be better to try a multivariate procedure such as multiple regression rather than a complicated univariate procedure such as Box-Jenkins, although this is not always the case.

†5.6 PREDICTION THEORY

Over the last thirty years or so, a general theory of linear prediction has been developed by Kolmogorov, Wiener (1949), Yaglom (1962) and Whittle (1963) among others. All these authors avoid the use of the word 'forecasting', although most of the univariate methods considered in Section 5.2 are in the general class of linear predictors. The theory of linear prediction has applications in control and communications engineering and is of considerable theoretical interest, but readers who wish to tackle the sort of forecasting problem we have been considering earlier in this chapter will find this literature less accessible than the other references. Here we will only give a brief introduction.

Two types of problem are often distinguished. In the first type of problem we have data up to time T, $\{x_T, x_{T-1}, \ldots\}$, and wish to predict the value of x_{T+m}. One approach is to use the predictor

$$\hat{x}_{T+m} = \Sigma c_j x_{T-j}$$

which is a linear function of the available data. The weights $\{c_j\}$ are chosen so

†This section may be omitted at first reading.

as to minimize the expected mean square prediction error, $E(x_{T+m}-\hat{x}_{T+m})^2$. This is often called the **prediction** problem (e.g. Cox and Miller, 1968), while Yaglom (1962) refers to it as the **extrapolation** problem and Whittle (1963) calls it **pure prediction**. As an example of the sort of result which has been obtained, Wiener (1949) has considered the problem of evaluating the weights $\{c_j\}$, so as to find the best linear predictor, when the ac.f. of the series $\{x_t\}$ is **known** and when the entire past sequence $\{x_t\}$ is known. It is interesting to compare this sort of approach with the forecasting techniques proposed earlier in this chapter. The Box-Jenkins approach, for example, also employs a linear predictor which will be optimal for a particular ARIMA process. But whereas Wiener says little about estimation, Box and Jenkins (1970) show how to find a linear predictor when the ac.f. has to be estimated.

The second type of problem arises when the process of interest, $s(t)$, called the **signal**, is contaminated by **noise**, $n(t)$, and we actually observe the process

$$y(t)=s(t)+n(t)$$

In some situations the noise is simply measurement error; in engineering applications the noise may be an interference process of some kind. The problem now is to separate the signal from the noise. Given measurements on $y(t)$ up to time T we may want to reconstruct the signal up to time T or alternatively make a prediction of $s(T+\tau)$. The problem of reconstructing the signal is often called **smoothing** or **filtering**. The problem of predicting the signal is also often called **filtering** (Yaglom, 1962; Cox and Miller, 1968), but is sometimes called **prediction** (Astrom, 1970). It is often assumed that the signal and noise processes are uncorrelated and that $s(t)$ and $n(t)$ have known ac.f.s.

It is clear that both the above types of problem are closely related to the control problem because, if we can predict how a process will behave, then we can adjust the process so that the achieved values are, in some sense, as close as possible to the target value. Further remarks on control theory must await a study of linear systems.

EXERCISES

5.1   For the MA(1) model given by

$$X_t=Z_t+\theta Z_{t-1}$$

show that $\hat{x}(N, 1)=\theta z_N$ and that $\hat{x}(N, k)=0$ for $k=2, 3, \ldots$.
Show that the variance of the k-steps-ahead forecast error is given by σ_Z^2 for $k=1$, and by $(1+\theta^2)\sigma_Z^2$ for $k\geq 2$, provided the true model is known. (In practice we would take $\hat{x}(N, 1)=\hat{\theta}\hat{z}_N$, where $\hat{\theta}$ is the least squares estimate of θ and $\hat{z}_N$ is the observed residual at time N.)

5.2   For the AR(1) model given by

$$X_t=\alpha X_{t-1}+Z_t$$

show that $\hat{x}(N, k) = \alpha^k x_N$ for $k = 1, 2, \ldots$. Also show that the variance of the k-steps-ahead forecast error is given by $(1 - \alpha^{2k})\sigma_Z^2/(1 - \alpha^2)$.

For the AR(1) model given by

$$X_t - \mu = \alpha(X_{t-1} - \mu) + Z_t$$

show that $\hat{x}(N, k) = \mu + \alpha^k(x_N - \mu)$ for $k = 1, 2, \ldots$. (In practice the least squares estimate of α would be substituted into the above formulae.)

5.3 Consider the SARIMA$(1, 0, 0)(0, 1, 1)_{12}$ model used as an example in Section 5.2.4. Show that

$$\hat{x}(N, 2) = x_{N-10} + \alpha^2(x_N - x_{N-12}) + \theta\alpha z_{N-11} + \theta z_{N-10}$$

5.4 For the SARIMA$(0, 0, 1)(1, 1, 0)_{12}$ model, find forecasts at time N for up to 12 steps ahead in terms of observations and estimated residuals up to time N.

5.5 For the model $(1 - B)(1 - 0.2B)X_t = (1 - 0.5B)Z_t$ in Exercise 3.12, find forecasts for one and two steps ahead, and show that a recursive expression for forecasts three or more steps ahead is given by

$$\hat{x}(N, k) = 1.2\hat{x}(N, k-1) - 0.2\hat{x}(N, k-2)$$

Find the variance of the one-, two- and three-steps-ahead forecast errors. If $z_N = 1$, $x_N = 4$, $x_{N-1} = 3$ and $\sigma_Z^2 = 2$, show that $\hat{x}(N, 2) = 3.64$ and that the standard error of the corresponding forecast error is 1.72.

5.6 Consider the ARIMA$(0, 1, 1)$ process

$$(1 - B)X_t = (1 - \theta B)Z_t$$

Show that $\hat{x}(N, 1) = x_N - \theta z_N$, and $\hat{x}(N, k) = \hat{x}(N, k-1)$ for $k \geq 2$. Express $\hat{x}(N, 1)$ in terms of x_N and $\hat{x}(N-1, 1)$ and show that this is equivalent to exponential smoothing. By considering the ψ weights of the process, show that the variance of the k-steps-ahead prediction error is $[1 + (k-1)(1 - \theta)^2]\sigma_Z^2$.

6
Stationary processes in the frequency domain

6.1 INTRODUCTION

In Chapter 3 we described several types of stationary stochastic process, placing emphasis on the autocovariance (or autocorrelation) function which is the natural tool for considering the evolution of a process through time. In this chapter we introduce a complementary function called the **spectral density function**, which is the natural tool for considering the frequency properties of a time series. Inference regarding the spectral density function is called an analysis in the **frequency domain**.

Some statisticians initially have difficulty in understanding the frequency approach, but the advantages of frequency methods are widely appreciated in such fields as electrical engineering, geophysics and meteorology. These advantages will become apparent in the next few chapters.

We shall confine ourselves to real-valued processes. Many authors consider the more general problem of complex-valued processes, and this results in some gain of mathematical conciseness. But, in my view, the reader is more likely to understand an approach restricted to real-valued processes. The vast majority of practical problems are covered by this approach.

6.2 THE SPECTRAL DISTRIBUTION FUNCTION

In order to introduce the idea of a spectral density function, we must first consider a function called the **spectral distribution function**. The approach adopted is heuristic and not mathematically rigorous, but will, hopefully, give the reader a better understanding of the subject than a more theoretical approach.

Suppose we suspect that a time series contains a periodic sinusoidal component with a known wavelength. Then a natural model is

$$X_t = R \cos(\omega t + \theta) + Z_t \qquad (6.1)$$

where ω is called the **frequency** of the sinusoidal variation, R is called the **amplitude** of the variation, θ is called the **phase**, and Z_t denotes some stationary random series. Note that the angle $(\omega t + \theta)$ is usually measured in units called radians, where π radians $= 180°$. Since ω is the number of radians per unit time it is sometimes called the **angular** frequency, but in keeping with most authors we call ω the frequency. However some authors, notably Jenkins and Watts (1968), refer to frequency as $f = \omega/2\pi$, the number of cycles per unit time, and this form of frequency is much easier to interpret from a physical point of view. We usually use the angular frequency ω in mathematical formulae for conciseness, but will often use the frequency $f = \omega/2\pi$ for the interpretation of data. The period of a sinusoidal cycle, called the **wavelength**, is clearly $1/f$ or $2\pi/\omega$. An example of a sinusoidal function is shown in Figure 6.1. There $f = 1/6$ and the wavelength is 6.

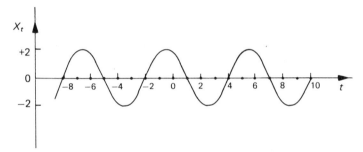

Figure 6.1   A graph of $R \cos(\omega t + \theta)$ with $R = 2$, $\omega = \pi/3$ and $\theta = \pi/6$.

Model (6.1) is a very simple model, but in practice the variation in a time series may be caused by variation at several different frequencies. For example, sales figures may contain weekly, monthly, yearly and other cyclical variation. In other words the data show variation at high, medium and low frequencies. It is natural therefore to generalize (6.1) to

$$X_t = \sum_{j=1}^{k} R_j \cos(\omega_j t + \theta_j) + Z_t \qquad (6.2)$$

where R_j is the amplitude at frequency ω_j.

The reader will notice that models (6.1) and (6.2) are **not** stationary if R, θ, $\{R_j\}$ and $\{\theta_j\}$ are fixed constants because $E(X_t)$ will change with time. In order to apply the theory of stationary processes to models like (6.2), it is customary to assume that $\{R_j\}$ are (uncorrelated) random variables with mean zero, or that $\{\theta_j\}$ are random variables with a uniform distribution on $(0, 2\pi)$, which are fixed for a single realization of the process (see Section 3.5 and Exercise 3.14). This is something of a 'mathematical trick', but it does enable us to treat time series containing one or more deterministic sinusoidal components as stationary series.

Since $\cos(\omega t + \theta) = \cos \omega t \cos \theta - \sin \omega t \sin \theta$, model (6.2) can be expressed as a sum of sine and cosine terms in the form

$$X_t = \sum_{j=1}^{k} (a_j \cos \omega_j t + b_j \sin \omega_j t) + Z_t \tag{6.3}$$

where $a_j = R_j \cos \theta_j$ and $b_j = - R_j \sin \theta_j$.

But we may now ask why there should only be a finite number of frequencies involved in model (6.2) or (6.3). In fact, letting $k \to \infty$, the work of Wiener and others has shown that any discrete stationary process measured at unit intervals may be represented in the form

$$X_t = \int_0^{\pi} \cos \omega t \; du(\omega) + \int_0^{\pi} \sin \omega t \; dv(\omega) \tag{6.4}$$

where $u(\omega)$, $v(\omega)$ are uncorrelated continuous processes with orthogonal increments (see Section 3.4.8) which are defined for all ω in the range $(0, \pi)$. Equation (6.4) is called the **spectral representation** of the process; it involves stochastic integrals, which require considerable mathematical skill to handle properly. It is intuitively more helpful to ignore these mathematical problems and simply regard X_t as a linear combination of orthogonal sinusoidal terms. Thus the derivation of the spectral representation will not be considered here (see for example Cox and Miller, 1968, Chapter 8).

The reader may wonder why the upper limits of the integrals in (6.4) are π rather than ∞. For a continuous process the upper limits would indeed be ∞, but for a discrete process measured at unit intervals of time there is no loss of generality in restricting ω to the range $(0, \pi)$, since

$$\cos[(\omega + k\pi)t] = \begin{cases} \cos \omega t & k, t \text{ integers with } k \text{ even} \\ \cos(\pi - \omega)t & k, t \text{ integers with } k \text{ odd} \end{cases}$$

and so variation at frequencies higher than π cannot be distinguished from variation at a corresponding frequency in $(0, \pi)$. The frequency $\omega = \pi$ is called the **Nyquist frequency**. We will say more about this in Section 7.2.1. For a discrete process measured at equal intervals of time of length Δt, the Nyquist frequency is $\pi / \Delta t$. In the next two sections we consider discrete processes measured at unit intervals of time, but the arguments carry over to discrete processes measured at intervals Δt if we replace π by $\pi / \Delta t$.

The main point of introducing the spectral representation (6.4) is to show that every frequency in the range $(0, \pi)$ may contribute to the variation of the process. However, the processes $u(\omega)$ and $v(\omega)$ in (6.4) are of little direct practical interest. Instead we introduce a function $F(\omega)$ called the (power) spectral distribution function, which arises from a theorem (e.g. Bartlett, 1966, Section 6.1), called the Wiener-Khintchine theorem, named after N. Wiener and A. Y. Khintchine. As applied to real-valued processes, this theorem says that, for any stationary stochastic process with autocovariance function $\gamma(k)$, there exists a monotonically increasing function $F(\omega)$ such that

$$\gamma(k) = \int_0^\pi \cos \omega k \, dF(\omega) \tag{6.5}$$

Equation (6.5) is called the spectral representation of the autocovariance function, and involves a type of integral (called Stieltjes) which may be unfamiliar to some readers. It can however be shown that the function $F(\omega)$ has a direct physical interpretation: it is the contribution to the variance of the series which is accounted for by frequencies in the range $(0, \omega)$. It is most important to understand this physical interpretation of $F(\omega)$. There is no variation at negative frequencies, so that

$$F(\omega) = 0 \qquad \text{for } \omega < 0$$

For a discrete process measured at unit intervals of time, the highest possible frequency is π and so all the variation is accounted for by frequencies less than π. Thus

$$F(\pi) = \text{Var}(X_t) = \sigma_X^2$$

This last result also comes directly from (6.5) with $k = 0$, when

$$\gamma(0) = \sigma_X^2 = \int_0^\pi dF(\omega) = F(\pi)$$

In between $\omega = 0$ and $\omega = \pi$, $F(\omega)$ is monotonically increasing.

If the process contains a deterministic sinusoidal component at frequency ω_0, say $R \cos(\omega_0 t + \theta)$ where R is a constant and θ is uniformly distributed on $(0, 2\pi)$, then there will be a step increase in $F(\omega)$ at ω_0 equal to $E[R^2 \cos^2(\omega_0 t + \theta)] = \frac{1}{2} R^2$.

As $F(\omega)$ is monotonic, it can be decomposed into two functions, $F_1(\omega)$ and $F_2(\omega)$, such that

$$F(\omega) = F_1(\omega) + F_2(\omega) \tag{6.6}$$

where $F_1(\omega)$ is a non-decreasing continuous function and $F_2(\omega)$ is a non-decreasing step function. This decomposition usually corresponds to the Wold decomposition, with $F_1(\omega)$ relating to the purely indeterministic component of the process and $F_2(\omega)$ relating to the deterministic component. We shall be mainly concerned with purely indeterministic processes, where $F_2(\omega) \equiv 0$, so that $F(\omega)$ is a continuous function on $(0, \pi)$.

The adjective 'power', which is sometimes prefixed to 'spectral distribution function', derives from the engineer's use of the word in connection with the passage of an electric current through a resistance. For a sinusoidal input, the power is directly proportional to the squared amplitude of the oscillation. For a more general input, the power spectral distribution function describes how the power is distributed with respect to frequency. In the case of a time series, the variance may be regarded as the total power.

Note that some authors use a normalized form of $F(\omega)$ given by

$$F^*(\omega) = F(\omega)/\sigma_X^2 \qquad (6.7)$$

Thus $F^*(\omega)$ is the **proportion** of variance accounted for by frequencies in the range $(0, \omega)$. Since $F^*(\pi) = 1$, and $F^*(\omega)$ is monotonically increasing, $F^*(\omega)$ has similar properties to a cumulative distribution function.

6.3   THE SPECTRAL DENSITY FUNCTION

For a purely indeterministic discrete stationary process, the spectral distribution function is a continuous (monotone bounded) function in $(0, \pi)$, and may therefore be differentiated with respect to ω in $(0, \pi)$. (Strictly speaking, $F(\omega)$ may not be differentiable on a set of measure zero, but this is of no practical importance.) We will denote the derivative by $f(\omega)$, so that

$$f(\omega) = \frac{\mathrm{d}F(\omega)}{\mathrm{d}\omega} \qquad (6.8)$$

This is the (power) spectral density function. The term 'spectral density function' is often shortened to **spectrum**, and the adjective 'power' is sometimes omitted.

When $f(\omega)$ exists, equation (6.5) can be expressed in the form

$$\gamma(k) = \int_0^\pi \cos \omega k f(\omega)\, \mathrm{d}\omega \qquad (6.9)$$

This is an ordinary (Riemann) integral and therefore much easier to handle. Putting $k = 0$, we have

$$\gamma(0) = \sigma_X^2 = \int_0^\pi f(\omega)\, \mathrm{d}\omega = F(\pi) \qquad (6.10)$$

The physical meaning of the spectrum is that $f(\omega)\,\mathrm{d}\omega$ represents the contribution to variance of components with frequencies in the range $(\omega, \omega + \mathrm{d}\omega)$. When the spectrum is drawn, equation (6.10) indicates that the total area underneath the curve is equal to the variance of the process. A peak in the spectrum indicates an important contribution to variance at frequencies in the appropriate interval. An example of a spectrum is shown in Figure 6.2, together with the corresponding normalized spectral distribution function.

It is important to realize that the autocovariance function (acv.f.) and the power spectral density function are equivalent ways of describing a stationary stochastic process. From a practical point of view, they are complementary to each other. Both functions contain the same information but express it in

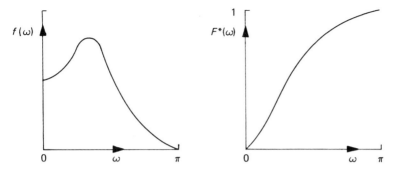

Figure 6.2 An example of a spectrum, together with the corresponding normalized spectral distribution function.

different ways. In some situations a time-domain approach based on the acv.f. is more useful, while in other situations a frequency-domain approach is preferable.

Equation (6.9) expresses $\gamma(k)$ in terms of $f(\omega)$ as a cosine transform. The inverse relationahip (see Appendix A) is given by

$$f(\omega) = \frac{1}{\pi} \sum_{k=-\infty}^{\infty} \gamma(k)e^{-i\omega k} \qquad (6.11)$$

so that the spectrum is the **Fourier transform** of the autocovariance function. Since $\gamma(k)$ is an even function, (6.11) is often written in the equivalent form

$$f(\omega) = \frac{1}{\pi}\left[\gamma(0) + 2\sum_{k=1}^{\infty} \gamma(k)\cos \omega k\right] \qquad (6.12)$$

Note that if we try to apply (6.12) to a process containing a deterministic component at frequency ω_0, then $\Sigma\gamma(k)\cos \omega_0 k$ will not converge, since $F(\omega)$ is not differentiable at ω_0 and so $f(\omega_0)$ is not defined.

The reader should note that several other definitions of the spectrum are given in the literature, most of which differ from (6.12) by a constant multiple and by the range of definition of $f(\omega)$. The most popular approach is to define the spectrum in the range $(-\pi, \pi)$ by

$$f(\omega) = \frac{1}{2\pi} \sum_{k=-\infty}^{\infty} \gamma(k)e^{-i\omega k} \qquad (6.13)$$

whose inverse relationship (see Appendix A) is

$$\gamma(k) = \int_{-\pi}^{\pi} e^{i\omega k}f(\omega)\, d\omega \qquad (6.14)$$

Jenkins and Watts (1968) use these equations, except that they take $f = \omega/2\pi$ as

the frequency variable (see equations (A.3) and (A.4)). Equations (6.13) and
(6.14), which form a Fourier transform pair, are the more usual form of the
Wiener-Khintchine relations. The formulation is slightly more general in that
it can be applied to complex-valued time series. But for real time series we find
that $f(\omega)$ is an even function, and then we need only consider $f(\omega)$ for $\omega > 0$. In
my experience the introduction of negative frequencies, while having certain
mathematical advantages, serves only to confuse the student. As we are
concerned only with real-valued processes, we prefer (6.11) defined on $(0, \pi)$.

   It is sometimes useful to use a normalized form of the spectral density
function, given by

$$f^*(\omega) = f(\omega)/\sigma_X^2 = \frac{\mathrm{d}F^*(\omega)}{\mathrm{d}\omega} \qquad (6.15)$$

This is the derivative of the normalized spectral distribution function (see
equation (6.7)). Then we find that $f^*(\omega)$ is the Fourier transform of the
autocorrelation function, namely

$$f^*(\omega) = \frac{1}{\pi}\left[1 + 2\sum_{k=1}^{\infty} \rho(k) \cos \omega k\right] \qquad (6.16)$$

and that $f^*(\omega)$ is the **proportion** of variance in the interval $(\omega, \omega + \mathrm{d}\omega)$.
Kendall, Stuart and Ord (1983, equation 47.20) define the spectral density
function in the range $(0, \pi)$ in terms of the autocorrelation function but omit
the constant $1/\pi$ from equation (6.16). This makes it more difficult to give the
function a physical interpretation. Instead they introduce an intensity function
which corresponds to our power spectrum.

6.4   THE SPECTRUM OF A CONTINUOUS PROCESS

For a continuous purely indeterministic stationary process $X(t)$, the
autocovariance function $\gamma(\tau)$ is defined for all τ and the (power) spectral
density function $f(\omega)$ is defined for all positive ω. The relationship between
these functions is very similar to that in the discrete case except that there is no
upper bound to the frequency. We have

$$f(\omega) = \frac{1}{\pi}\int_{-\infty}^{\infty} \gamma(\tau)e^{-i\omega\tau}\,\mathrm{d}\tau$$

$$= \frac{2}{\pi}\int_{0}^{\infty} \gamma(\tau)\cos\omega\tau\,\mathrm{d}\tau \qquad (6.17)$$

for $0 < \omega < \infty$, with the inverse relationship

$$\gamma(\tau) = \int_{0}^{\infty} f(\omega)\cos\omega\tau\,\mathrm{d}\omega \qquad (6.18)$$

6.5 DERIVATION OF SELECTED SPECTRA

In this section we derive the spectral density functions of some simple but important stationary processes.

(a) *Purely random process*

A purely random process in discrete time, $\{Z_t\}$, is defined in Section 3.4.1. If $\mathrm{Var}(Z_t) = \sigma_Z^2$, then the acv.f. is given by

$$\gamma(k) = \begin{cases} \sigma_Z^2 & k = 0 \\ 0 & \text{otherwise} \end{cases}$$

so that the power spectral density function is given by

$$f(\omega) = \sigma_Z^2/\pi \qquad (6.19)$$

using (6.12). In other words the spectrum is constant in the range $(0, \pi)$.

We have already pointed out that a continuous white noise process is physically unrealizable. A process is regarded as a practical approximation to continuous white noise if its spectrum is substantially constant over the frequency band of interest, even if it then approaches zero at high frequency.

(b) *First-order moving average process*

The first-order MA process (see Section 3.4.3)

$$X_t = Z_t + \beta Z_{t-1}$$

has an ac.f, given by

$$\rho(k) = \begin{cases} 1 & k = 0 \\ \beta/(1 + \beta^2) & k = \pm 1 \\ 0 & \text{otherwise} \end{cases}$$

So, using (6.16), the normalized spectral density function is given by

$$f^*(\omega) = \frac{1}{\pi} [1 + (2\beta \cos \omega)/(1 + \beta^2)] \qquad (6.20)$$

for $0 < \omega < \pi$. The power spectral density function is then

$$f(\omega) = \sigma_X^2 f^*(\omega)$$

where $\sigma_X^2 = (1 + \beta^2)\sigma_Z^2$.

The shape of the spectrum depends on the value of β. When $\beta > 0$ the power is concentrated at low frequencies, giving what is called a low-frequency spectrum; if $\beta < 0$ the power is concentrated at high frequencies, giving a high-frequency spectrum. Examples are shown in Figure 6.3.

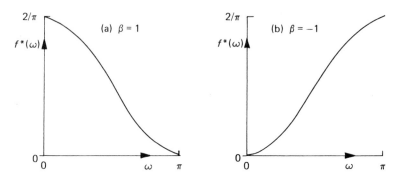

Figure 6.3 Two examples of spectra of first-order moving average processes with, (a) $\beta=1$; (b) $\beta=-1$.

(c) *First-order autoregressive process*

The first-order AR process (see Section 3.4.4)

$$X_t = \alpha X_{t-1} + Z_t \tag{6.21}$$

has an acv.f. given by

$$\gamma(k) = \sigma_X^2 \, \alpha^{|k|} \qquad k = 0, \pm 1, \pm 2, \ldots$$

The power spectral density function is then, using (6.11),

$$f(\omega) = \frac{\sigma_X^2}{\pi}\left(1 + \sum_{k=1}^{\infty} \alpha^k e^{-ik\omega} + \sum_{k=1}^{\infty} \alpha^k e^{ik\omega}\right)$$

$$= \frac{\sigma_X^2}{\pi}\left(1 + \frac{\alpha e^{-i\omega}}{1-\alpha e^{-i\omega}} + \frac{\alpha e^{i\omega}}{1-\alpha e^{i\omega}}\right)$$

which after some algebra gives

$$f(\omega) = \sigma_X^2 (1-\alpha^2)/[\pi(1-2\alpha \cos \omega + \alpha^2)] \tag{6.22}$$

$$= \sigma_Z^2/\pi(1-2\alpha \cos \omega + \alpha^2) \tag{6.23}$$

since $\sigma_Z^2 = \sigma_X^2(1-\alpha^2)$.

The shape of the spectrum depends on the value of α. When $\alpha > 0$ the power is concentrated at low frequencies, while if $\alpha < 0$ the power is concentrated at high frequencies. Examples are shown in Figure 6.4.

It is hoped that the reader finds the shapes of the spectra in Figure 6.4 intuitively reasonable. For example if α is negative then it is clear from (6.21) that values of X_t will tend to oscillate, and rapid oscillations correspond to high-frequency variation.

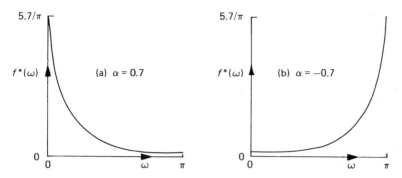

Figure 6.4   Two examples of spectra of first-order autoregressive processes with, (a) $\alpha=0.7$; (b) $\alpha=-0.7$.

(d) *Higher-order autoregressive processes*

It can be shown (e.g. Jenkins and Watts, 1968) that the spectrum of a second-order AR process with parameters α_1, α_2 is given by

$$f(\omega)=\sigma_Z^2/\pi[1+\alpha_1^2+\alpha_2^2-2\alpha_1(1-\alpha_2)\cos\omega-2\alpha_2\cos 2\omega]$$

for $0<\omega<\pi$. The shape of the spectrum depends on the values of α_1 and α_2. It is possible to get a high-frequency spectrum, a low-frequency spectrum, a spectrum with a peak between 0 and π, or a spectrum with a minimum between 0 and π.

For higher-order AR processes, one can get spectra with several peaks or troughs.

(e) *A deterministic sinusoidal perturbation*

Suppose that

$$X_t=\cos(\omega_0 t+\theta) \tag{6.24}$$

where ω_0 is a constant in $(0, \pi)$ and θ is a random variable which is uniformly distributed on $(0, 2\pi)$. As explained in Section 3.5, θ is fixed for a single realization of the process and (6.24) defines a purely deterministic process.

The acv.f. of the process is given by

$$\gamma(k)=\tfrac{1}{2}\cos\omega_0 k$$

which we note does **not** tend to zero as k increases. This is a feature of most deterministic processes.

From (6.24) it is obvious that all the 'power' of the process is concentrated at the frequency ω_0. Now $\mathrm{Var}(X_t)=E(X_t^2)=\tfrac{1}{2}$, so that the power spectral distribution function is given by

$$F(\omega)=\begin{cases}0 & \omega<\omega_0\\ \tfrac{1}{2} & \omega\geq\omega_0\end{cases}$$

Since this is a step function, it has no derivative at ω_0 and so the spectrum is not defined at ω_0. If we nevertheless try to use equation (6.12) to obtain the spectrum as the Fourier transform of the acv.f. then we find that

$$f(\omega)=0 \qquad \omega \neq \omega_0$$

but that $\Sigma \gamma(k) \cos \omega k$ does not converge at $\omega = \omega_0$.

(f) *A mixture*

Our final example contains a mixture of deterministic and stochastic components, namely

$$X_t = \cos(\omega_0 t + \theta) + Z_t$$

where ω_0, θ are as defined in example (e) above, and $\{Z_t\}$ is a purely random process with mean zero and variance σ_Z^2. Then we find that the acv.f. is given by

$$\gamma(k) = \begin{cases} \frac{1}{2} + \sigma_Z^2 & k = 0 \\ \frac{1}{2} \cos \omega_0 k & k = \pm 1, \pm 2, \dots \end{cases}$$

Again note that $\gamma(k)$ does **not** tend to zero as k increases because X_t contains a deterministic component.

We can obtain the power spectral distribution function by using (6.6), since the deterministic component $\cos(\omega_0 t + \theta)$ has a distribution function

$$F_1(\omega) = \begin{cases} 0 & \omega < \omega_0 \\ \frac{1}{2} & \omega \geqslant \omega_0 \end{cases}$$

while the stochastic component Z_t has a distribution function

$$F_2(\omega) = \sigma_Z^2 \omega / \pi \qquad 0 < \omega < \pi$$

on integrating (6.19). Thus the overall spectral distribution function is given by

$$F(\omega) = \begin{cases} \sigma_Z^2 \omega / \pi & 0 < \omega < \omega_0 \\ \frac{1}{2} + \sigma_Z^2 \omega / \pi & \omega_0 \leqslant \omega < \pi \end{cases}$$

As in example (e), the power spectrum is not defined at $\omega = \omega_0$.

EXERCISES

In the following questions $\{Z_t\}$ denotes a purely random process, mean zero and variance σ_Z^2.

6.1 Find (a) the power spectral density function, (b) the normalized spectral density function of the first-order AR process

$$X_t = \lambda X_{t-1} + Z_t$$

(Note: (a) is covered in the text, but see if you can do it without looking it up.)

6.2 Find the power spectral density functions of the following MA processes:

(a) $X_t = Z_t + Z_{t-1} + Z_{t-2}$
(b) $X_t = Z_t + 0.5Z_{t-1} - 0.3Z_{t-2}$

6.3 Show that the second-order MA process

$$X_t = \mu + Z_t + 0.8Z_{t-1} + 0.5Z_{t-2}$$

is second-order stationary, where μ denotes a constant. Find the acv.f. and ac.f. of $\{X_t\}$ and show that its normalized spectral density function is given by

$$f^*(\omega) = (1 + 1.27 \cos \omega + 0.53 \cos 2\omega)/\pi \qquad 0 < \omega < \pi$$

6.4 A stationary time series $(X_t; t = \ldots, -1, 0, +1, \ldots)$ has normalized spectral density function

$$f^*(\omega) = 2(\pi - \omega)/\pi^2 \qquad 0 < \omega < \pi$$

Show that its ac.f. is given by

$$\rho(k) = \begin{cases} 1 & k = 0 \\ (2/\pi k)^2 & k \text{ odd} \\ 0 & k \text{ even } (\neq 0) \end{cases}$$

6.5 A two-state Markov process may be set up as follown. Alpha particles fiuom a radioactive source are used to trigger a flip-flop device which takes the states $+1$ and -1 alternately. The times t_i at which changes occur constitute a Poisson process, with mean event rate λ. Let $X(t)$ denote the state variable at time t. If the process is started at $t = 0$ with $P[X(0) = 1] = P[X(0) = -1] = \frac{1}{2}$, show that the process is second-order stationary, with autocorrelation function

$$\rho(u) = e^{-2\lambda|u|} \qquad -\infty < u < \infty$$

and spectral density function

$$f(\omega) = 4\lambda/[\pi(4\lambda^2 + \omega^2)] \qquad 0 < \omega < \infty$$

6.6 Show that if $\{X_t\}$ and $\{Y_t\}$ are independent, stationary processes with power spectral density functions $f_x(\omega)$ and $f_y(\omega)$, then $\{V_t\} = \{X_t + Y_t\}$ is also stationary with power spectral density function $f_v(\omega) = f_x(\omega) + f_y(\omega)$. If

$$V_t = X_t + Y_t$$

where

$$X_t = \alpha X_{t-1} + W_t \qquad -1 < \alpha < +1$$

and $\{Y_t\}, \{W_t\}$ are independent purely random processes with zero mean and common variance σ^2, show that the power spectral density function of $\{V_t\}$ is given by

$$f_v(\omega) = \sigma^2(2 - 2\alpha \cos \omega + \alpha^2)/\pi(1 - 2\alpha \cos \omega + \alpha^2) \qquad 0 < \omega < \pi$$

6.7   Show that the normalized spectral density function of the ARMA(1, 1) process

$$X_t = \alpha X_{t-1} + Z_t + \beta Z_{t-1}$$

is given by

$$f^*(\omega) = \frac{1}{\pi}[1 + 2\rho(1)(\cos \omega - \alpha)/(1 - 2\alpha \cos \omega + \alpha^2)] \qquad 0 < \omega < \pi$$

(Hint: Use the results in Exercise 3.11. Note that an easier way of finding the power spectral density function of ARMA processes is given in Chapter 9: see Exercise 9.5)

7
Spectral analysis

Spectral analysis is the name given to methods of **estimating** the spectral density function, or spectrum, of a given time series.

In the last century, research workers such as A. Schuster were essentially concerned with looking for 'hidden periodicities' in data, but spectral analysis as we know it today is mainly concerned with estimating the spectrum over the whole range of frequencies. The techniques are now widely used by many scientists, particularly in electrical engineering, physics, meteorology and marine science.

We are mainly concerned with purely indeterministic processes, which have a continuous spectrum, but the techniques can also be used for deterministic processes to pick out periodic components in the presence of noise.

7.1 FOURIER ANALYSIS

Spectral analysis is essentially a modification of **Fourier** analysis so as to make it suitable for stochastic rather than deterministic functions of time. Fourier analysis (e.g. Priestley, 1981) is basically concerned with approximating a function by a sum of sine and cosine terms, called the Fourier series representation. Suppose that a function $f(t)$ is defined on $(-\pi, \pi]$ and satisfies the so-called Dirichlet conditions. These conditions ensure that $f(t)$ is reasonably 'well behaved', i.e. that over the range $(-\pi, \pi]$, $f(t)$ is absolutely integrable, has a finite number of discontinuities, and has a finite number of maxima and minima. Then $f(t)$ may be approximated by the Fourier series

$$\frac{a_0}{2} + \sum_{r=1}^{k} (a_r \cos rt + b_r \sin rt)$$

where

$$a_0 = \frac{1}{\pi} \int_{-\pi}^{\pi} f(t) \, dt$$

$$a_r = \frac{1}{\pi} \int_{-\pi}^{\pi} f(t) \cos rt \, dt \qquad r = 1, 2, \ldots$$

$$b_r = \frac{1}{\pi} \int_{-\pi}^{\pi} f(t) \sin rt \, dt \qquad r = 1, 2, \ldots$$

It can be shown that this Fourier series converges to $f(t)$ as $k \to \infty$ except at points of discontinuity, where it converges to half-way up the step change. Mathematicians say that this is the average of the limit from below and the limit from above, and write it as $\frac{1}{2}[f(t-0)+f(t+0)]$.

In order to apply Fourier analysis to discrete time series, we need to consider the Fourier series representation of $f(t)$ when $f(t)$ is defined only on the integers $1, 2, \ldots, N$. Rather than write down the formula, we demonstrate that the required Fourier series emerges naturally by considering a simple sinusoidal model.

7.2   A SIMPLE SINUSOIDAL MODEL

Suppose we suspect that a given time series, with observations made at unit time intervals, contains a deterministic sinusoidal component at frequency ω together with a random error term. So we will consider the model

$$X_t = \mu + \alpha \cos \omega t + \beta \sin \omega t + Z_t \tag{7.1}$$

where Z_t denotes a purely random process, and μ, α, β are parameters to be estimated from the data.

The observations will be denoted by $(x_1, x_2, \ldots, x_N)$. The algebra in the next few sections is somewhat simplified if we confine ourselves to the case where N is **even**. There is no real difficulty in extending the results to the case where N is odd (e.g. Anderson, 1971), and indeed many of the later estimation formulae apply for both odd and even N, but some results require one to consider odd N and even N separately. Thus, if N happens to be odd and a spectral analysis is required, it can make things simpler to remove the first observation so as to make N even. If N is reasonably large, little information is lost.

Model (7.1) can be represented in matrix notation by

$$E(X) = A\theta$$

where

$$X^T = (X_1, \ldots, X_N)$$

$$\theta^T = (\mu, \alpha, \beta)$$

$$A = \begin{pmatrix} 1 & \cos \omega & \sin \omega \\ 1 & \cos 2\omega & \sin 2\omega \\ \cdots & \cdots & \cdots \\ 1 & \cos N\omega & \sin N\omega \end{pmatrix}$$

As this model is linear in the parameters μ, α and β, it is an example of a general linear model. In that case the least squares estimate of θ, which minimizes $\sum_{t=1}^{N} (x_t - \mu - \alpha \cos \omega t - \beta \sin \omega t)^2$, is 'well known' to be

$$\hat{\theta} = (A^T A)^{-1} A^T \mathbf{x}$$

where

$$\mathbf{x}^T = (x_1, \ldots, x_N)$$

Now the highest frequency we can fit to the data is the Nyquist frequency, given by $\omega = \pi$, while the lowest frequency we can reasonably fit completes one cycle in the whole length of the time series (see Section 7.2.1). By equating the cycle length $2\pi/\omega$ to N, we find that this lowest frequency is given by $2\pi/N$. The least squares estimates are particularly simple if ω is restricted to one of the values

$$\omega_p = 2\pi p/N \qquad p = 1, \ldots, N/2$$

as $(A^T A)$ then turns out to be a diagonal matrix in view of the following 'well-known' trigonometric results (all summations are for $t = 1$ to N):

$$\Sigma \cos \omega_p t = \Sigma \sin \omega_p t = 0 \tag{7.2}$$

$$\Sigma \cos \omega_p t \cos \omega_q t = \begin{cases} 0 & p \neq q \\ N & p = q = N/2 \\ N/2 & p = q \neq N/2 \end{cases} \tag{7.3}$$

$$\Sigma \sin \omega_p t \sin \omega_q t = \begin{cases} 0 & p \neq q \\ 0 & p = q = N/2 \\ N/2 & p = q \neq N/2 \end{cases} \tag{7.4}$$

$$\Sigma \cos \omega_p t \sin \omega_q t = 0 \qquad \text{for all } p, q \tag{7.5}$$

With $(A^T A)$ diagonal we can easily find $\hat{\theta}$ For ω_p, $p \neq N/2$, we find (Exercise 7.2)

$$\hat{\mu} = \Sigma x_t/N = \bar{x}$$
$$\hat{\alpha} = 2\Sigma x_t \cos(\omega_p t)/N \tag{7.6}$$
$$\hat{\beta} = 2\Sigma x_t \sin(\omega_p t)/N$$

If $p = N/2$ we ignore the term in $\beta \sin \omega t$, which is zero for all t, and find

$$\hat{\mu} = \bar{x}$$
$$\hat{\alpha} = \Sigma x_t (-1)^t/N \tag{7.7}$$

Model (7.1) is essentially the one used in the last century to search for hidden periodicities, but this model has now gone out of fashion. However it can still be useful if we have reason to suspect that a time series does contain a deterministic periodic component at a known frequency and we wish to isolate this component (e.g. Bloomfield, 1976, Chapter 2; Pocock, 1974).

Readers who are familiar with the analysis of variance technique will see that the total corrected sum of squared deviations, namely

$$\sum_{t=1}^{N} (x_t - \bar{x})^2$$

can be partitioned into two components which are the residual sum of squares and the sum of squares 'explained' by the periodic component at frequency ω_p. This latter component is given by

$$\sum_{t=1}^{N} (\hat{\alpha} \cos \omega_p t + \hat{\beta} \sin \omega_p t)^2$$

which after some algebra (Exercise 7.2) can be shown to be

$$\begin{array}{ll} (\hat{\alpha}^2 + \hat{\beta}^2)N/2 & p \neq N/2 \\ \hat{\alpha}^2 N & p = N/2 \end{array} \tag{7.8}$$

using (7.2)–(7.5).

7.2.1 The Nyquist frequency

In Chapter 6 we pointed out that for a discrete process measured at unit intervals there is no loss of generality in restricting the spectral distribution function to the range $(0, \pi)$. We now demonstrate that the upper bound π, called the Nyquist frequency, is indeed the highest frequency about which we can get meaningful information from a set of data.

First we will give a more general form for the Nyquist frequency. If observations are taken at equal intervals of length Δt, then the Nyquist (angular) frequency is $\omega_N = \pi/\Delta t$. The equivalent frequency expressed in cycles per unit time is $f_N = \omega_N/2\pi = 1/2\Delta t$.

Consider the following example. Suppose that temperature readings are taken every day in a certain town at noon. It is clear that these observations will tell us nothing about temperature variation **within** a day. In particular they will not tell us if nights are hotter or cooler than days. With only one observation per day, the Nyquist frequency is $\omega_N = \pi$ radians per day or $f_N = \frac{1}{2}$ cycle per day (1 cycle per two days). This is lower than the frequencies which correspond to variation within a day. For example variation with a wavelength of one day has (angular) frequency $\omega = 2\pi$ radians per day or $f = 1$ cycle per day. In order to get information about variation within a day at these higher frequencies, we must increase the sampling rate and take two or more observations per day.

A similar example is provided by yearly sales figures. These will obviously give no information about any seasonal effects, whereas monthly or quarterly observations **will** give information about seasonality.

Finally, we make a comment about the lowest frequency we can fit to a set of data. If we had just six months of temperature readings from winter to summer it would not be clear if there was an upward trend in the observations or if winters are colder than summers. However, with one year's data it **would** become clear that winters are colder than summers. Thus if we are interested in variation at the low frequency of 1 cycle per year, then we must have at least one year's data. Thus the lower the frequency we are interested in, the longer the time period over which we need to take measurements, whereas the higher the frequency we are interested in, the more frequently must we take observations.

7.3 PERIODOGRAM ANALYSIS

Early attempts at discovering hidden periodicities in a given time series basically consisted of repeating the analysis of Section 7.2 at all the frequencies $2\pi/N$, $4\pi/N$, ..., π. In view of (7.3)–(7.5) the different terms are orthogonal and we end up with the finite Fourier series representation of the $\{x_t\}$, namely

$$x_t = a_0 + \sum_{p=1}^{(N/2)-1} [a_p \cos(2\pi pt/N) + b_p \sin(2\pi pt/N)] + a_{N/2} \cos \pi t$$

$$t = 1, 2, \ldots, N \quad (7.9)$$

where the coefficients $\{a_p, b_p\}$ are of the same form as equations (7.6) and (7.7), namely

$$a_0 = \bar{x}$$

$$a_{N/2} = \Sigma(-1)^t x_t / N$$

$$\left.\begin{array}{l} a_p = 2[\Sigma x_t \cos(2\pi pt/N)]/N \\ b_p = 2[\Sigma x_t \sin(2\pi pt/N)]/N \end{array}\right\} \quad p = 1, \ldots, (N/2) - 1$$

$$(7.10)$$

An analysis along these lines is sometimes called a Fourier analysis or a **harmonic** analysis. The Fourier series representation (7.9) has N parameters to describe N observations and so can be made to fit the data exactly (just as a polynomial of degree $N-1$ involving N parameters can be found which goes exactly through N observations in polynomial regression). This explains why there is no error term in (7.9) in contrast to (7.1). Also note that there is no term in sin πt in (7.9) as sin πt is zero for all integer t.

It is worth stressing that the Fourier series coefficients (7.10) at a given frequency ω are exactly the same as the least squares estimates for model (7.1).

The overall effect of the Fourier analysis of the data is to partition the variability of the series into components at frequencies $2\pi/N$, $4\pi/N$, ..., π. The component at frequency $\omega_p = 2\pi p/N$ is often called the pth harmonic. For $p \neq N/2$, it is often useful to write the pth harmonic in the equivalent form

$$a_p \cos \omega_p t + b_p \sin \omega_p t = R_p \cos(\omega_p t + \phi_p) \qquad (7.11)$$

where

$$R_p = \sqrt{(a_p^2 + b_p^2)} \qquad (7.12)$$

is the **amplitude** of the pth harmonic, and

$$\phi_p = \tan^{-1}(-b_p/a_p) \qquad (7.13)$$

is the **phase** of the pth harmonic.

We have already noted in Section 7.2 that, for $p \neq N/2$, the contribution of the pth harmonic to the total sum of squares is given by $N(a_p^2 + b_p^2)/2$. Using (7.12), this is equal to $NR_p^2/2$. Extending this result using (7.2)–(7.5) and (7.9), we have, after some algebra, that (Exercise 7.3)

$$\sum_{t=1}^{N} (x_t - \bar{x})^2 = N \sum_{p=1}^{(N/2)-1} R_p^2/2 + Na_{N/2}^2 \qquad (7.14)$$

Dividing through by N we have

$$\Sigma(x_t - \bar{x})^2/N = \sum_{p=1}^{(N/2)-1} R_p^2/2 + a_{N/2}^2 \qquad (7.15)$$

which is known as Parseval's theorem. The left-hand side of (7.15) is effectively the variance of the observations, although the divisor is N rather than the more usual $(N-1)$. Thus $R_p^2/2$ is the contribution of the pth harmonic to the variance, and (7.15) shows how the total variance is partitioned.

If we plot $R_p^2/2$ against $\omega_p = 2\pi p/N$ we obtain a line spectrum. A different type of line spectrum occurs in the physical sciences when light from molecules in a gas discharge tube is viewed through a spectroscope. The light has energy at discrete frequencies and this energy can be seen as bright lines. But most time series have continuous spectra, and then it is inappropriate to plot a line spectrum. If we regard $R_p^2/2$ as the contribution to variance in the range $\omega_p \pm \pi/N$, we can plot a histogram whose height in the range $\omega_p \pm \pi/N$ is such that

$$R_p^2/2 = \text{area of histogram rectangle}$$

$$= \text{height of histogram} \times 2\pi/N$$

Thus the height of the histogram is given by

$$I(\omega_p) = NR_p^2/4\pi \qquad (7.16)$$

As usual, (7.16) does not apply for $p = N/2$; we may regard $a_{N/2}^2$ as the contribution to variance in the range $[\pi(N-1)/N, \pi]$ so that

$$I(\pi) = Na_{N/2}^2/\pi$$

The plot of $I(\omega)$ against ω is usually called the **periodogram** even though $I(\omega)$ is a function of frequency rather than period. Other authors define the periodogram in a slightly different way, as some other multiple of R_p^2. Hannan (1970, equation (3.8)) defines the periodogram in terms of complex numbers as

$$\frac{1}{2\pi N}\left|\sum_{t=1}^{N} x_t e^{it\omega}\right|^2$$

which is $\frac{1}{2} \times$ expression (7.16). Anderson (1971, Section 4.3.2) describes the graph of R_p^2 against the **period** N/p as the periodogram, and suggests the term **spectrogram** to describe the graph of R_p^2 against frequency. An advantage of definition (7.16) is that the total area under the periodogram is equal to the variance of the time series. Expression (7.16) may readily be calculated directly from the data by

$$I(\omega_p) = [(\Sigma x_t \cos 2\pi pt/N)^2 + (\Sigma x_t \sin 2\pi pt/N)^2]/N\pi \qquad (7.17)$$

Equation (7.17) also applies for $p = N/2$. Jenkins and Watts (1968) define a similar expression in terms of the variable $f = \omega/2\pi$, but call it the 'sample spectrum'.

The periodogram appears to be a natural way of estimating the power spectral density function, but we shall see that for a process with a **continuous** spectrum it provides a poor estimate and needs to be modified.

7.3.1 The relationship between the periodogram and the autocovariance function

The periodogram ordinate $I(\omega)$ and the autocovariance coefficient c_k are both quadratic forms of the data $\{x_t\}$. It is of interest to see how they are related. We will show that the periodogram is the finite Fourier transform of $\{c_k\}$.

Using (7.2) we may rewrite (7.17) for $p \neq N/2$ as

$$I(\omega_p) = \{[\Sigma(x_t - \bar{x}) \cos \omega_p t]^2 + [\Sigma(x_t - \bar{x}) \sin \omega_p t]^2\}/N\pi$$

$$= \sum_{s,t=1}^{N} (x_t - \bar{x})(x_s - \bar{x})(\cos \omega_p t \cos \omega_p s + \sin \omega_p t \sin \omega_p s)/N\pi$$

But (see (4.1))

$$\sum_{t=1}^{N-k} (x_t - \bar{x})(x_{t+k} - \bar{x})/N = c_k$$

and

$$\cos \omega_p t \cos \omega_p(t+k) + \sin \omega_p t \sin \omega_p(t+k)$$

$$= \cos \omega_p(t+k-t)$$

$$= \cos \omega_p k$$

so that

$$I(\omega_p) = (c_0 + 2 \sum_{k=1}^{N-1} c_k \cos \omega_p k)/\pi \tag{7.18}$$

$$= \sum_{k=-(N-1)}^{N-1} c_k e^{-i\omega_p k}/\pi \tag{7.19}$$

We recognize (7.19) as a finite Fourier transform (assuming that $c_k = 0$ for $|k| \geqslant N$).

7.3.2 Properties of the periodogram

When the periodogram is expressed in the form (7.18), it appears to be the 'obvious' estimate of the power spectrum

$$f(\omega) = (\gamma_0 + 2 \sum_{k=1}^{\infty} \gamma_k \cos \omega k)/\pi$$

simply replacing γ_k by its estimate c_k for values of k up to $(N-1)$, and putting subsequent estimates of γ_k equal to zero. But although we find

$$\underset{N \to \infty}{E} [I(\omega)] \to f(\omega) \tag{7.20}$$

so that the periodogram is asymptotically unbiased, we will see that the variance of $I(\omega)$ does not decrease as N increases. Thus $I(\omega)$ is **not a consistent** estimator for $f(\omega)$. An example of a periodogram is given in Figure 7.5(c), and it can be seen that the graph fluctuates wildly. The lack of consistency is perhaps not too surprising when one realizes that the Fourier series representation (7.9) requires one to evaluate N parameters from N observations however long the series. Thus in Section 7.4 we will consider alternative ways of estimating a power spectrum which are essentially ways of **smoothing** the periodogram.

We complete this section by proving that $I(\omega)$ is not a consistent estimator for $f(\omega)$ in the case where $(x_1, \ldots, x_N)$ are taken from a discrete purely random process, where the observations are independent $N(\mu, \sigma^2)$ variates. This result can be extended to other stationary processes with continuous spectra, but this will not be demonstrated here.

From (7.10) we see that a_p and b_p are linear combinations of normally distributed random variables and so will themselves be normally distributed. Using (7.2)–(7.4), it can be shown (Exercise 7.4) that a_p and b_p each have mean zero and variance $2\sigma^2/N$ for $p \neq N/2$. Furthermore we have

$$\text{Cov}(a_p, b_p) = 4 \, \text{Cov}[(\Sigma x_t \cos \omega_p t), (\Sigma x_t \sin \omega_p t)]/N^2$$

$$= 4\sigma^2 (\Sigma \cos \omega_p t \sin \omega_p t)/N^2$$

since the $\{x_t\}$ are independent. Thus, using (7.5), we see that a_p and b_p are uncorrelated. Since (a_p, b_p) are bivariate normal, zero correlation implies that a_p and b_p are independent. Now a result from distribution theory says that if Y_1, Y_2 are independent $N(0, 1)$ variables, then $(Y_1^2 + Y_2^2)$ has a χ^2 distribution with two degrees of freedom, which is written χ_2^2. Thus

$$\frac{N(a_p^2 + b_p^2)}{2\sigma^2} = \frac{I(\omega_p)2\pi}{\sigma^2}$$

is χ_2^2. Now the variance of a χ^2 distribution with v degrees of freedom is $2v$, so that

$$\mathrm{Var}[I(\omega_p)2\pi/\sigma^2] = 4$$

and

$$\mathrm{Var}[I(\omega_p)] = \sigma^4/\pi^2$$

As this variance is a constant, it does **not** tend to zero as $N \to \infty$, and hence $I(\omega_p)$ is not a consistent estimator for $f(\omega_p)$. Furthermore it can be shown that neighbouring periodogram ordinates are asymptotically independent, which further explains the very irregular form of an observed periodogram. Thus the periodogram needs to be modified in order to obtain a good estimate of a continuous spectrum.

7.4 SPECTRAL ANALYSIS: SOME CONSISTENT ESTIMATION PROCEDURES

This section describes several alternative procedures for carrying out a spectral analysis. The different methods will be compared in Section 7.6. Each method provides a **consistent** estimate of the (power) spectral density function, in contrast to the periodogram. But although the periodogram is itself an inconsistent estimate, we shall see that the procedures described in this section are essentially based on the periodogram by using some sort of smoothing procedure.

Throughout the section we will assume that any obvious trend or seasonal variation has been removed from the data. If this is not done, the results of the spectral analysis are likely to be dominated by these effects, making any other effects difficult or impossible to see. Trend will produce a peak at zero frequency, while seasonal variation produces peaks at the seasonal frequency and at integer multiples of the seasonal frequency. These integer multiples of the fundamental frequency are called **harmonics** (see Section 7.8). For a non-stationary series, the estimated spectrum can depend rather crucially on the method chosen to remove trend and seasonality.

The methods described in this chapter are essentially non-parametric in that no model is assumed *a priori*. An alternative parametric approach called autoregressive spectrum estimation will be introduced in Section 11.5.

7.4.1   Transforming the truncated autocovariance function

One type of estimation procedure consists of taking a Fourier transform of the truncated weighted sample autocovariance function. From equation (7.18), we have that the periodogram is the discrete Fourier transform of the complete sample autocovariance function. But it is clear that the precision of the c_k decreases as k increases, so that it would seem intuitively reasonable to give less weight to the values of c_k as k increases. An estimator which has this property is

$$\hat{f}(\omega) = \frac{1}{\pi} \left\{ \lambda_0 c_0 + 2 \sum_{k=1}^{M} \lambda_k c_k \cos \omega k \right\} \tag{7.21}$$

where $\{\lambda_k\}$ are a set of weights called the **lag window**, and $M(<N)$ is called the **truncation point**. Comparing (7.21) with (7.18) we see that values of c_k for $M < k < N$ are no longer used, while values of c_k for $k \leqslant M$ are weighted by a factor λ_k.

In order to use the above estimator, the reader must choose a suitable lag window and a suitable truncation point. The two best-known lag windows are as follows.

(a) *Tukey window*

$$\lambda_k = \frac{1}{2}\left(1 + \cos \frac{\pi k}{M}\right) \qquad k = 0, 1, \ldots, M$$

This window is also called the Tukey-Hanning or Blackman-Tukey window.

(b) *Parzen window*

$$\lambda_k = \begin{cases} 1 - 6\left(\dfrac{k}{M}\right)^2 + 6\left(\dfrac{k}{M}\right)^3 & 0 \leqslant k \leqslant M/2 \\ 2(1 - k/M)^3 & M/2 \leqslant k \leqslant M \end{cases}$$

These two windows are illustrated in Figure 7.1 with $M = 20$.

The Tukey and Parzen windows will give very much the same estimated spectrum for a given time series, although the Parzen window has a slight advantage in that it cannot give negative estimates. Many other lag windows have been suggested (see Hannan, 1970, Section 5.4), and 'window carpentry' was a popular research topic in the 1950s. Ways of comparing different windows will be discussed in Section 7.6. The well-known Bartlett window, with $\lambda_k = 1 - k/M$ for $k = 0, 1, \ldots, M$, is no longer used as its properties are

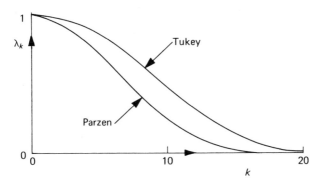

Figure 7.1   The Tukey and Parzen lag windows with $M = 20$.

inferior to the Tukey and Parzen windows. Neave (1972b) has suggested an alternative window which has superior properties but which is more complicated to use.

The choice of the truncation point M is rather difficult and little clear-cut advice is available in the literature. It has to be chosen subjectively so as to balance 'resolution' against 'variance'. The smaller the value of M, the smaller will be the variance of $\hat{f}(\omega)$ but the larger will be the bias. If M is too small, important features of $f(\omega)$ may be smoothed out, while if M is too large the behaviour of $\hat{f}(\omega)$ becomes more like that of the periodogram with erratic variation. Thus a compromise value must be chosen. A useful rough guide is to choose M to be about $2\sqrt{N}$, so that if for example N is 200, then M will be round about the value 28. This choice of M ensures the asymptotic situation that as $N \to \infty$, so also does $M \to \infty$ but in such a way that $M/N \to 0$. A somewhat larger value of M is required for the Parzen window than for the Tukey window. Jenkins and Watts (1968) suggest trying three different values of M. A low value will give an idea where the large peaks in $f(\omega)$ are, but the curve is likely to be too smooth. A high value is likely to produce a curve showing a large number of peaks, some of which may be spurious. A compromise can then be achieved with the third value of M. As Hannan (1970, p. 311) says, 'experience is the real teacher and that cannot be got from a book.'

In principle (7.21) may be evaluated at any value of ω in $(0, \pi)$, but it is usually evaluated at equal intervals at $\omega = \pi j/Q$ for $j = 0, 1, \ldots, Q$, where Q is chosen sufficiently large to show up all features of $\hat{f}(\omega)$. Often Q is chosen to be equal to M. The graph of $\hat{f}(\omega)$ against ω can then be plotted and examined. An example is given in Figure 7.5 for the data plotted in Figure 1.2 using the Tukey window with $M = 24$.

7.4.2 Hanning

This procedure, named after Julius Von Hann, is equivalent to the use of the Tukey window as described in Section 7.4.1, but adopts a different computational procedure. The estimated spectrum is calculated in two stages. First, a truncated unweighted cosine transform of the data is taken to give

$$\hat{f}_1(\omega) = \frac{1}{\pi}\left(c_0 + 2\sum_{k=1}^{M} c_k \cos \omega k\right) \tag{7.22}$$

This is the same as (7.21) except that the lag window is taken to be unity (i.e. $\lambda_k = 1$). The estimates given by (7.22) are calculated at $\omega = \pi j/M$ for $j = 0, 1, \ldots, M$. These estimates are then smoothed using the weights $(\frac{1}{4}, \frac{1}{2}, \frac{1}{4})$ to give the Hanning estimates

$$\hat{f}(\omega) = \tfrac{1}{4}\hat{f}_1(\omega - \pi/M) + \tfrac{1}{2}\hat{f}_1(\omega) + \tfrac{1}{4}\hat{f}_1(\omega + \pi/M) \tag{7.23}$$

at $\omega = \pi j/M$ for $j = 1, 2, \ldots, (M-1)$. At zero frequency, and at the Nyquist frequency π, we take

$$\hat{f}(0) = \tfrac{1}{2}[\hat{f}_1(0) + \hat{f}_1(\pi/M)]$$
$$\hat{f}(\pi) = \tfrac{1}{2}[\hat{f}_1(\pi) + \hat{f}_1(\pi(M-1)/M)]$$

It is easily demonstrated that this procedure is equivalent to the use of the Tukey window. Substituting (7.22) into (7.23) we find

$$\hat{f}(\omega) = \frac{1}{\pi}\left\{c_o + 2\sum_{k=1}^{M} c_k[\tfrac{1}{4}\cos(\omega - \pi/M)k + \tfrac{1}{2}\cos \omega k + \tfrac{1}{4}\cos(\omega + \pi/M)k]\right\}$$

But

$$\cos(\omega - \pi/M)k + \cos(\omega + \pi/M)k = 2\cos \omega k \cos(\pi k/M)$$

and comparing with (7.21) we see that the lag window is indeed the Tukey window.

There is relatively little difference in the computational efficiency of Hanning and the straightforward use of the Tukey window. Both methods yield the same estimates and so it matters little which of the two procedures is used in practice.

7.4.3 Hamming

This technique is very similar to Hanning and has a very similar title, which sometimes leads to confusion. In fact Hamming is named after a quite different person, namely R. W. Hamming. The technique is nearly identical to Hanning except that the weights $(\frac{1}{4}, \frac{1}{2}, \frac{1}{4})$ in (7.23) are changed to $(0.23, 0.54, 0.23)$. At the frequencies $\omega = 0$ and $\omega = \pi$, the weights are 0.54 and 0.46. The procedure gives similar estimates to those produced by Hanning.

7.4.4 Smoothing the periodogram

The methods of Sections 7.4.1–7.4.3 are based on transforming the sample autocovariance function. An alternative type of approach is to smooth the periodogram by simply grouping the periodogram ordinates in sets of size m and finding their average value. This approach is based on a suggestion by P. J. Daniell in 1946. Then we find

$$\hat{f}(\omega) = \frac{1}{m} \sum_j I(\omega_j) \qquad (7.24)$$

where $\omega_j = 2\pi j/N$ and j varies over m consecutive integers so that the ω_j are symmetric about ω. In order to estimate $f(\omega)$ at $\omega = 0$ and $\omega = \pi$, (7.24) has to be modified in an obvious way, treating the periodogram as being symmetric about 0 and π. Then, taking m to be odd with $m^* = (m-1)/2$, we have

$$\hat{f}(0) = 2 \sum_{j=1}^{m^*} I(2\pi j/N)/m$$

assuming $I(0) = 0$. We also have

$$\hat{f}(\pi) = \left[I(\pi) + 2 \sum_{j=1}^{m^*} I(\pi - 2\pi j/N) \right] \bigg/ m$$

Now we know that the periodogram is asymptotically unbiased but inconsistent for the true spectrum. Since neighbouring periodogram ordinates are asymptotically uncorrelated, it is clear that the variance of (7.24) will be of order $1/m$. It is also clear that the estimator (7.24) may be biased since

$$E[\hat{f}(\omega)] \simeq \frac{1}{m} \sum_j f(\omega_j)$$

which is equal to $f(\omega)$ only if the spectrum is linear over the interval. However, the bias will be unimportant provided that $f(\omega)$ is a reasonably smooth function and m is not too large compared with N.

Thus the choice of m is rather like the choice of the truncation point M in Section 7.4.1 in that it has to be chosen so as to balance resolution against variance, although the effects are in opposite directions. The larger the value of m the smaller will be the variance of the resulting estimate but the larger will be the bias, and if m is too large then interesting features of $f(\omega)$, such as peaks, may be smoothed out. As N increases, so we can allow m to increase.

There seems to be relatively little advice in the literature on the choice of m. It seems advisable to try several values, in the region of $N/40$. A high value should give some idea where the large peaks in $f(\omega)$ are, but the curve is likely to be too smooth. A low value is likely to produce a curve showing a large number of peaks, some of which may be spurious. A compromise can then be made.

Although the procedure described in this section is computationally quite different from that of Section 7.4.1, there are in fact close links between the two procedures. In Section 7.3.1 we derived the relationship between the periodogram and the sample autocovariance function, and if we substitute equation (7.18) into (7.24) we can express the estimate $\hat{f}(\omega)$ in terms of the sample autocovariance function in a similar form to equation (7.21). We find, after some algebra (Exercise 7.5), that the truncation point is $(N-1)$ and the lag window is given by

$$\lambda_k - \begin{cases} 1 & k=0 \\ \dfrac{sin(m\pi k/N)}{m\,\sin(\pi k/N)} & k=1, 2, \ldots, N-1 \end{cases}$$

This lag window works reasonably well, but has the undesirable property that it does not tend to zero as k tends to N. This illustrates that a sudden cut-off in the frequency domain can give rise to 'nasty' effects in the time domain, and vice versa. Because of this, it is worth noting that it is possible to smooth the periodogram by a variety of non-uniform averaging procedures, such as Hanning, but they will not be considered here.

Historically, the smoothed periodogram was not much used until recent years because it apparently requires much more computational effort than the procedure of Section 7.4.1. Calculating the periodogram using equation (7.17) at ω_p for $p=1, 2, \ldots, N/2$ would require about N^2 arithmetic operations (each one a multiplication and an addition), whereas using equation (7.21) fewer than MN operations are required to calculate the $\{c_k\}$ so that the total number of operations is only of order $M(N+M)$ if $Q=M$. Two factors have led to the increasing use of the smoothed periodogram. First, the advent of high-speed computers means that it is no longer necessary to restrict oneself to the method requiring the fewest calculations. The second factor has been the rediscovery of a technique called the fast Fourier transform which can speed up the computation of the periodogram quite considerably. This technique will now be described.

7.4.5   The fast Fourier transform

This technique substantially reduces the time required to perform a Fourier analysis on a computer, and is also more accurate. The title is usually abbreviated to FFT and we will use this abbreviation. (But note that Hannan, 1970, uses this abbreviation to denote finite Fourier transform.)

A history of the FFT is described by Cooley, Lewis and Welch (1967), the ideas going back to the early 1900s. But it was the work of J. W. Cooley,

J. W. Tukey and G. Sande about 1965 which first stimulated the application of the technique to time-series analysis. Much of the subsequent work was published in *IEEE Transactions on Audio and Electroacoustics* (since succeeded by *IEEE Transactions on Acoustics, Speech and Signal Processing*). We will only give a broad outline of the technique here. For further details, see for example Bendat and Piersol (1986), Bloomfield (1976) and Priestley (1981).

The basic idea of the FFT can be illustrated in the case when N can be factorized in the form $N=rs$. If we assume that N is even, then at least one of the factors, say r, will be even. Using complex numbers for mathematical simplicity, the Fourier coefficients from equation (7.10) are given by

$$a_p + ib_p = 2[\Sigma x_t\, e^{2\pi i pt/N}]/N \qquad (7.25)$$

for $p=0, 1, 2, \ldots, (N/2)-1$. For mathematical convenience we denote the observations by $x_0, x_1, \ldots, x_{N-1}$, so that the summation in (7.25) is from $t=0$ to $N-1$. Now we can write t in the form

$$t = rt_1 + t_0$$

where $t_1 = 0, 1, \ldots, s-1$, and $t_0 = 0, 1, \ldots, r-1$, as t goes from 0 to $N-1$, in view of the fact that $N=rs$. Similarly we can decompose p in the form

$$p = sp_1 + p_0$$

where $p_1 = 0, 1, \ldots, (r/2)-1$, and $p_0 = 0, 1, \ldots, s-1$, as p goes from 0 to $(N/2)-1$. Then the summation in (7.25) may be written

$$\sum_{t_0=0}^{r-1} e^{2\pi i p t_0/N} \sum_{t_1=0}^{s-1} x_t\, e^{2\pi i prt_1/N}$$

But

$$e^{2\pi i prt_1/N} = e^{2\pi i(sp_1 + p_0)rt_1/N} = e^{2\pi i p_0 rt_1 N}$$

since $e^{2\pi i sp_1 rt_1/N} = e^{2\pi i p_1 t_1} = 1$ for all p_1, t_1. Thus $\sum_{t_1=0}^{s-1} x_t\, e^{2\pi i prt_1/N}$ does not depend on p_1 and is therefore a function of t_0 and p_0 only, say $A(p_0, t_0)$. Then (7.25) may be written

$$a_p + ib_p = 2\left[\sum_{t_0=0}^{r-1} A(p_0, t_0)\, e^{2\pi i p t_0/N}\right]/N$$

Now there are rs functions of type $A(p_0, t_0)$ to be calculated, each requiring s complex multiplications and additions. Then the $a_p + ib_p$ may be calculated with $r^2s/2$ complex multiplications and additions, giving a grand total of $rs(s+r/2) = N(s+r/2)$ calculations instead of the $N^2/2$ calculations required to use (7.25) directly.

Much bigger reductions in computing can be made by an extension of the above procedure when N is highly composite (i.e. has many small factors). In particular, if N is of the form 2^n then we find that the number of operations is of order Nn (or $N\log_2 N$) instead of N^2. Substantial gains can also be made when N has several factors (e.g. $N=2^p 3^q 5^r \ldots$).

In practice it is unlikely that N will be of a simple form such as 2^n, although it may be possible to make N highly composite by omitting a few observations. More generally we can add zeros to the (mean-corrected) sample record so as to **increase** the value of N until it is a suitable integer. Then a procedure called **tapering** or data windowing (e.g. Bloomfield, 1976; Priestley, 1981) is sometimes recommended to avoid a discontinuity at the end of the data, though its use is now controversial. Suppose for example that we happen to have 382 observations. This value of N is **not** highly composite and we might proceed as follows:

(a) Remove any linear trend from the data, and keep the residuals (which should have mean zero) for subsequent analysis. If there is no trend, simply subtract the overall mean from each observation.

(b) Apply a linear taper to about 5% of the data at each end. In this example, if we denote the detrended mean-corrected data by $x_0, x_1, \ldots, x_{381}$, then the tapered series is given by

$$x_t^* = \begin{cases} (t+1)x_t/20 & t=0, 1, \ldots, 18 \\ (382-t)x_t/20 & t=363, \ldots, 381 \\ x_t & t=19, 20, \ldots, 362 \end{cases}$$

(c) Add $512-382=130$ zeros at one end of the tapered series, so that $N=512=2^9$.

(d) Carry out an FFT on the data, calculate the Fourier coefficients $a_p + ib_p$, and average the values of $(a_p^2 + b_p^2)$ in groups of about 10.

In fact with N as low as 382, the computational advantage of the FFT is limited and we could equally well calculate the periodogram directly, which avoids the need for tapering and adding zeros. The FFT really comes into its own for $N>1000$.

Jenkins and Watts (1968) give two reasons why they think the case for using the FFT in spectral analysis is not strong. First, they say that fast computers are more than adequate for carrying out a spectral analysis by traditional methods. This is certainly true for say $N<1000$, but perhaps not for several thousand observations. Secondly, Jenkins and Watts say that the autocorrelation function is an invaluable intermediate stage in spectral analysis. This is also true, but does not mean that the FFT is useless because it can be quicker to calculate the sample autocovariance function by performing **two** FFTs (e.g. Priestley, 1981, Section 7.6). We can compute the Fourier coefficients (a_p, b_p) with an FFT of the mean-corrected data at $\omega_p = 2\pi p/N$ for $p=0, 1, \ldots, N-1$ and not for $p=0, 1, \ldots, N/2$ as we usually do. The extra coefficients are redundant for real-valued processes since $a_{N-k}=a_k$ and $b_{N-k}=-b_k$. We then compute $R_p^2=a_p^2+b_p^2$ and fast Fourier retransform the sequence (R_p^2) to get the mean lagged products. With $N=2^{12}=4096$, for example, R. Fenton found that it took over three times as long to calculate one-sixth of the $\{c_k\}$ directly than to calculate **all** the $\{c_k\}$ using two FFTs. In using the FFT for this purpose, one has to be careful to add enough zeros to the data (**without** tapering) to make sure that one gets non-circular sums of lagged products, as defined by equation (4.1) and used throughout this book. Circular coefficients result if zeros are not added where, for example, the circular autocovariance coefficient at lag 1 is

$$c_1^* = \left[\sum_{t=1}^{N}(x_t-\bar{x})(x_{t+1}-\bar{x})\right]/N$$

where x_{N+1} is taken equal to x_1. If $x_1=\bar{x}$, the circular and non-circular coefficients at lag 1 are the same. To calculate all the autocovariance coefficients of a set of N observations one adds N zeros, to make $2N$ 'observations' in all.

7.5   CONFIDENCE INTERVALS FOR THE SPECTRUM

The methods of Section 7.4 all produce **point** estimates of the spectral density function at different frequencies. In this section we show how to find confidence intervals for the spectrum at different frequencies.

In Section 7.3.2 we showed that a white noise process, with constant spectrum $f(\omega)=\sigma^2/\pi$, has a periodogram ordinate $I(\omega)$ at frequency ω which is such that $2I(\omega)/f(\omega)$ is distributed as χ_2^2. A more general result is given by Jenkins and Watts (1968, Section 6.4.2) for the estimator of Section 7.4.1, namely

$$\hat{f}(\omega)=\left[\sum_{k=-M}^{M}\lambda_k c_k \cos \omega k\right]/\pi$$

which is that asymptotically $v\hat{f}(\omega)/f(\omega)$ is approximately distributed as χ_v^2, where

$$v=2N\left/\sum_{k=-M}^{M}\lambda_k^2\right.\qquad\qquad(7.26)$$

is called the number of degrees of freedom of the lag window. Then

$$P(\chi_{v,1-\alpha/2}^2 < v\hat{f}(\omega)/f(\omega) < \chi_{v,\alpha/2}^2)=1-\alpha$$

so that the $100(1-\alpha)\%$ confidence interval for $f(\omega)$ is given by

$$\frac{v\hat{f}(\omega)}{\chi_{v,\alpha/2}^2}\quad\text{to}\quad\frac{v\hat{f}(\omega)}{\chi_{v,1-\alpha/2}^2}$$

The degrees of freedom for the Tukey and Parzen windows turn out to be $2.67N/M$ and $3.71N/M$ respectively. When smoothing the periodogram in groups of size m, it is clear that the result will have $2m$ degrees of freedom and there is no need to apply equation (7.26). In fact equation (7.26) does not work for the periodogram when expressed in the form (7.18), as noted by Hannan (1970, p. 281).

The confidence intervals given in this section are asymptotic results. Neave (1972a) has shown that these results are also quite accurate for short series.

7.6 A COMPARISON OF DIFFERENT ESTIMATION PROCEDURES

Several factors need to be considered when comparing the different estimation procedures which were described in Section 7.4. These include such practical considerations as computing time and the availability of computer programs. We begin by considering the theoretical properties of the different procedures. Other comparative discussions are given by Jenkins and Watts (1968), Neave (1972b), Bloomfield (1976) and Priestley (1981, Section 7.5).

It is useful to introduce a function called the **spectral window** or **kernel**, which is the Fourier transform of the lag window. If we define the lag window λ_k to be zero for $k > M$, and to be symmetric so that $\lambda_{-k} = \lambda_k$, then the spectral window is given by

$$K(\omega) = \frac{1}{2\pi} \sum_{k=-\infty}^{\infty} \lambda_k e^{-ik\omega} \tag{7.27}$$

for $(-\pi < \omega < \pi)$. This has the inverse relation

$$\lambda_k = \int_{-\pi}^{\pi} K(\omega) e^{i\omega k}\, d\omega \tag{7.28}$$

All the estimation procedures for the spectrum can be put in the general

$$
\begin{aligned}
\hat{f}(\omega_0) &= \frac{1}{\pi} \sum_{k=-N+1}^{N-1} \lambda_k c_k e^{-i\omega_0 k} \\
&= \frac{1}{\pi} \Sigma \left[\int_{-\pi}^{\pi} K(\omega) e^{i\omega k}\, d\omega \right] c_k e^{-i\omega_0 k} \\
&= \frac{1}{\pi} \int_{-\pi}^{\pi} K(\omega) \left[\Sigma c_k e^{ik(\omega - \omega_0)} \right] d\omega \\
&= \int_{-\pi}^{\pi} K(\omega) I(\omega_0 - \omega)\, d\omega
\end{aligned}
\tag{7.29}
$$

form using equation (7.19). Equation (7.29) shows that all the estimation procedures are essentially smoothing the periodogram using the weight function $K(\omega)$. The value of the lag window at lag zero is usually specified to be one, so that from (7.28) we have

$$\lambda_0 = 1 = \int_{-\pi}^{\pi} K(\omega)\, d\omega$$

which is a desirable property for a smoothing function.

Taking expectations in equation (7.29) we have asymptotically that

$$E[\hat{f}(\omega_0)] = \int_{-\pi}^{\pi} K(\omega) f(\omega_0 - \omega)\, d\omega \qquad (7.30)$$

Thus the spectral window is a weight function expressing the contribution of the spectral density function at each frequency to the expectation of $\hat{f}(\omega_0)$. The name 'window' arises from the fact that $K(\omega)$ determines the part of the periodogram which is 'seen' by the estimator.

Examples of the spectral windows for the three commonest methods of spectral analysis are shown in Figure 7.2, which is adapted from Jones (1965, Figure 5). Taking $N = 1000$, the spectral window for the smoothed periodogram with $m = 20$ is shown as line A. The other two windows are the Parzen and Tukey windows, denoted by lines B and C. The values of the truncation point M were chosen to be 93 for the Parzen window and 67 for the Tukey window. These values of M were chosen so that all three windows gave estimators with equal variance. Formulae for variances will be given later in this section.

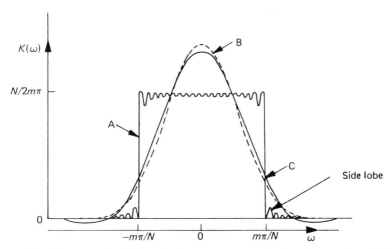

Figure 7.2   The spectral windows for three common methods of spectral analysis: A, smoothed periodogram $(m = 20)$; B, Parzen $(M = 93)$; C, Tukey $(M = 67)$; all with $N = 1000$.

Inspecting Figure 7.2, we see that the Parzen and Tukey windows look very similar, although the Parzen window has the advantage of being non-negative and of having smaller side-lobes. The shape of the periodogram window is quite different. It is approximately rectangular with a sharp cut-off and is close to the 'ideal' band-pass filter, which would be exactly rectangular but which is unattainable in practice. The periodogram window also has the advantage of being non-negative.

In comparing different windows, we also want to consider bias, variance and bandwidth. We will not derive formulae for the bias produced by the different procedures. It is clear from equation (7.30) that the wider the window, the larger will be the bias. In particular it is clear that all the smoothing procedures will tend to lower peaks and raise troughs.

As regards variance, we have from Section 7.5 that $v\hat{f}(\omega)/f(\omega)$ is approximately distributed as χ^2_v, where $v = 2m$, $3.71\,N/M$, and $8N/3M$ for the smoothed periodogram, Parzen and Tukey windows, respectively. Since

$$\text{Var}(\chi^2_v) = 2v$$

and

$$\text{Var}[v\hat{f}(\omega)/f(\omega)] = v^2\,\text{Var}[\hat{f}(\omega)/f(\omega)]$$

we find $\text{Var}[\hat{f}(\omega)/f(\omega)]$ turns out to be $1/m$, $2M/3.71N$, and $3M/4N$ for the three windows. Equating these expressions gives the values of M chosen for Figure 7.2.

Finally let us introduce the term **bandwidth**, which roughly speaking is the width of the spectral window. Various definitions are given in the literature, but we shall adopt that used by Jenkins and Watts (1968), namely the width of the 'ideal' rectangular window which would give an estimator with the same variance. The window of the smoothed periodogram is so close to being rectangular for m 'large' that it is clear from Figure 7.2 that the bandwidth will be $2\pi m/N$. The bandwidths for the Parzen and Tukey windows turn out to be $2\pi(1.86/M)$ and $8\pi/3M$ respectively. When plotting a graph of an estimated spectrum, it is a good idea to indicate the bandwidth which has been used.

The choice of bandwidth, or equivalently the choice of m or M, is an important step in spectral analysis. For the Parzen and Tukey windows, the bandwidth is inversely proportional to M (see Figure 7.3). As M gets larger, the window gets narrower and the bias gets smaller but the variance of the resulting estimator gets larger. In fact the variance is inversely proportional to the bandwidth. For the smoothed periodogram, the bandwidth is directly proportional to m. For the unsmoothed periodogram, with $m = 1$, the window is very tall and narrow giving an estimator with large variance as we have already shown. All in all, the choice of bandwidth is rather like the choice of class interval when constructing a histogram.

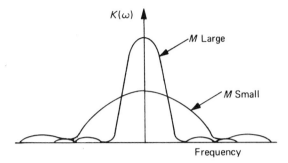

Figure 7.3   Spectral windows for different values of M.

We will now summarize the comparative merits of the three main estimation procedures. From the preceding discussion, it appears that the smoothed periodogram has superior theoretical properties in that its spectral window is approximately rectangular. From a computing point of view, it can be much slower for large N unless the FFT is used. If the FFT is used, however, it can be much quicker and it is also possible to calculate the autocorrelation function using two FFTs. For small N, computing time is relatively unimportant. Regarding computer programs, it is much easier to write a program for the Parzen or Tukey windows, but programs and algorithms for the FFT are becoming readily available. Thus the use of the smoothed periodogram is becoming more general.

7.7   ANALYSING A CONTINUOUS TIME SERIES

We have so far been concerned with the spectral analysis of discrete time series. But time series are sometimes recorded as a continuous trace, as for example air temperature, the moisture content of tobacco emerging from a processing plant, and humidity. For series which contain components at very high frequencies, such as those arising in acoustics and speech processing, it may be possible to analyse them mechanically using tuned filters, but the more usual procedure is to **digitize** the series by reading off the values of the trace at discrete intervals. If values are taken at equal time intervals of length Δt, we have converted a continuous time series into a standard discrete time series and can use the methods already described.

In sampling a continuous time series, the main question is how to choose the sampling interval Δt. It is clear that sampling leads to some loss of information and that this loss gets worse as Δt increases. However, it is expensive to make Δt very small and so a compromise value must be sought.

For the sampled series, the Nyquist frequency is $\pi/\Delta t$ radians per unit time, and we can get no information about variation at higher frequencies. Thus we clearly want to choose Δt so that variation in the continuous series is negligible at frequencies higher than $\pi/\Delta t$. In fact most measuring instruments are **band-limited** in that they do not respond to frequencies higher than a certain maximum frequency. If this maximum frequency is known or can be guessed, then the choice of Δt is straightforward.

If Δt is chosen to be too large, then a phenomenon called **aliasing** may occur. This can be illustrated by the following theorem.

Theorem 7.1   A continuous time series, with spectrum $f_c(\omega)$ for $0<\omega<\infty$, is sampled at equal intervals of length Δt. The resulting discrete time series has spectrum $f_d(\omega)$ defined over $0<\omega<\pi/\Delta t$. Then $f_d(\omega)$ and $f_c(\omega)$ are related by

$$f_d(\omega) = \sum_{s=0}^{\infty} f_c(\omega+2\pi s/\Delta t) + \sum_{s=1}^{\infty} f_c(-\omega+2\pi s/\Delta t) \tag{7.31}$$

Proof   The proof will be given for the case $\Delta t=1$. The extension to other values of Δt is straightforward.

The acv.f.s of the continuous and sampled series will be denoted by $\gamma(\tau)$ and γ_k. It is clear that when τ takes an integer value, say k, then

$$\gamma(k)=\gamma_k \tag{7.32}$$

Now from (6.9) we have

$$\gamma_k = \int_0^\pi f_d(\omega)\cos \omega k \, d\omega$$

while from (6.18) we have

$$\gamma(\tau) = \int_0^\infty f_c(\omega)\cos \omega\tau \, d\omega$$

Thus, using (7.32), we have

$$\int_0^\pi f_d(\omega)\cos \omega k \, d\omega = \int_0^\infty f_c(\omega)\cos \omega k \, d\omega$$

for $k=0, \pm 1, \pm 2, \ldots$. Now

$$\int_0^\infty f_c(\omega)\cos \omega k \, d\omega = \sum_{s=0}^{\infty} \int_{2\pi s}^{2\pi(s+1)} f_c(\omega)\cos \omega k \, d\omega$$

$$= \sum_{s=0}^{\infty} \int_0^{2\pi} f_c(\omega+2\pi s)\cos \omega k \, d\omega$$

$$= \sum_{s=0}^{\infty} \int_0^\pi \{f_c(\omega+2\pi s)+f_c[2\pi(s+1)-\omega]\}\cos \omega k \, d\omega$$

$$= \int_0^\pi \left\{\sum_{s=0}^{\infty} f_c(\omega+2\pi s) + \sum_{s=1}^{\infty} f_c(2\pi s-\omega)\right\}\cos \omega k \, d\omega$$

and the result follows.

The implications of this theorem are now considered. First, we note that if the continuous series contains no variation at frequencies above the Nyquist frequency, so that $f_c(\omega)=0$ for $\omega > \pi/\Delta t$, then $f_d(\omega)=f_c(\omega)$. In this case no information is lost by sampling. But more generally, the effect of sampling will be that variation at frequencies above the Nyquist frequency will be 'folded back' and produce an effect at a frequency lower than the Nyquist frequency in $f_d(\omega)$. If we denote the Nyquist frequency, $\pi/\Delta t$, by ω_N, then the frequencies ω, $2\omega_N - \omega$, $2\omega_N + \omega$, $4\omega_N - \omega$, ... are called **aliases** of one another. Variation at all these frequencies in the continuous series will appear as variation at frequency ω in the sampled series.

From a practical point of view, aliasing will cause trouble unless Δt is chosen so that $f_c(\omega) \simeq 0$ for $\omega > \pi/\Delta t$. If we have no advance knowledge about $f_c(\omega)$ then we can guesstimate a value for Δt. If the resulting estimate of $f_d(\omega)$ approaches zero near the Nyquist frequency $\pi/\Delta t$, then our choice of Δt was almost certainly sufficiently small. But if $f_d(\omega)$ does not approach zero near the Nyquist frequency, then it is probably wise to try a smaller value of Δt. Alternatively one can filter the continuous series to remove the high-frequency components if one is interested in the low-frequency components.

7.8 DISCUSSION

Spectral analysis can be a useful exploratory diagnostic tool in the analysis of many types of time series. In this section we discuss how the estimated spectrum should be interpreted, when it is likely to be most useful and when it is likely to be least useful. We also discuss some of the practical problems arising in spectral analysis.

We begin this discussion with an example to give the reader some feel for the sorts of spectrum shape that may arise. Figure 7.4 shows four sections of trace, labelled A, B, C and D, which were produced by four different processes (generated in a control engineering laboratory). The figure also shows the corresponding long-run spectra, labelled J, K, L and M, but these are given in **random** order. Note that the four traces use the same scale, the length produced in one second being shown on trace D. The four spectra are plotted using the same linear scales. The peak in spectrum L is at 15 cycles per second (or 15 Hz). The reader is invited to decide which series goes with which spectrum before reading on.

Trace A is much smoother than the other three traces. Its spectrum is therefore concentrated at low frequency and is actually spectrum M. The other three spectra are much harder to distinguish. Trace B is somewhat smoother than C or D and corresponds to spectrum K, which 'cuts off' at a lower frequency than J or L. Trace C corresponds to spectrum J, while trace D contains a deterministic sinusoidal component at 15 cycles per second which contributes 20% of the total power. Thus D corresponds to spectrum L.

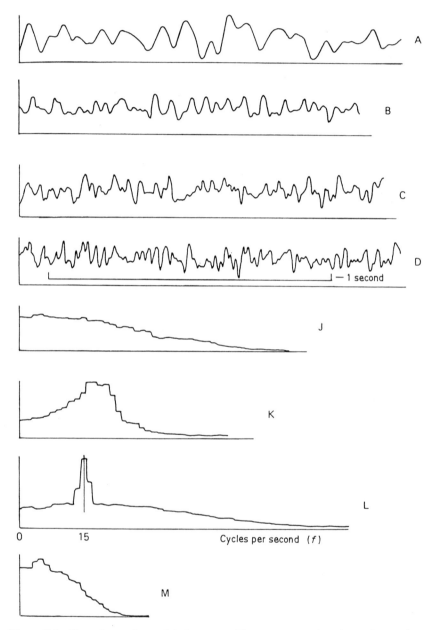

Figure 7.4 Four time series and their spectra. The spectra are given in random order.

From a visual inspection of traces C and D, it is difficult or impossible to decide which goes with spectrum L. For this type of data, spectral analysis is invaluable in assessing the frequency properties. The reader may find it surprising that the deterministic component in trace D is so hard to see. In contrast the regular seasonal variation in air temperature at Recife given in Figure 1.2 is quite obvious from a visual inspection of the time plot, but there the deterministic component accounts for 85% of the total variation. A spectral analysis of air temperature at Recife yields the spectrum shown in Figure 7.5(a) with a large peak at a frequency of one cycle per year. But here the spectral analysis is not really necessary as the seasonal effect is obvious anyway. In fact if one has a series containing an obvious trend or seasonality, then such variation should be removed from the data **before** carrying out a spectral analysis, as any other effects will be relatively small and are unlikely to be visible in the spectrum of the raw data. Figure 7.5(b) shows the spectrum of the Recife air temperature data when the seasonal variation has been removed. The variance is concentrated at low frequencies, indicating either a trend which is not apparent in Figure 1.2, or short-term correlation as in a first-order AR process with a positive coefficient (cf. Figure 6.4(a)).

Removing trend and seasonality is the simplest form of a general procedure called **prewhitening**. It is easier to estimate the spectrum of a series which has a relatively flat spectrum. Prewhitening consists of making a linear transformation of the raw data so as to achieve a smoother spectrum, estimating the spectrum of the transformed data, and then using the transfer function of the linear transformation to estimate the spectrum of the raw data (see Chapter 9 and Anderson, 1971, p. 546). But this procedure requires some prior knowledge of the spectral shape and is not often used except for removing trend and seasonality.

Having estimated the spectrum of a given time series, how do we interpret the results? There are various features to look for. First, are there any peaks in the spectrum? If so, why? Secondly, is the spectrum large at low frequency, indicating possible non-stationarity in the mean? Thirdly, what is the general shape of the spectrum? The typical shape of the power spectrum of an economic variable is shown in Figure 7.6, and the implications of this shape are discussed by Granger (1966). There are exceptions. Granger and Hughes (1971) found a peak at a frequency of 1 cycle per 13 years when they analysed Beveridge's yearly wheat price index series. But this series is much longer than most economic series, and the results have, in any case, been queried (Akaike, 1978).

The general shape of the spectrum may occasionally be helpful in indicating an appropriate parametric model, but it is not generally used in this way. The spectrum is not, for example, used in the Box-Jenkins procedure for indentifying an appropriate ARIMA process (though Hannan, 1970, Section 6.5 has suggested that it might be). Spectral analysis is essentially a

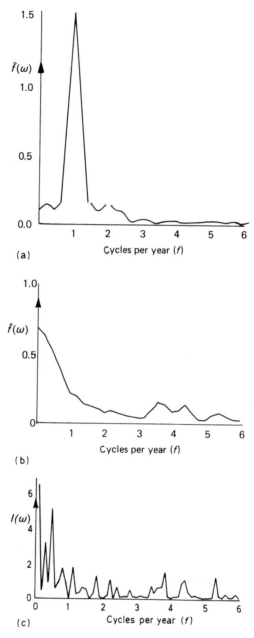

Figure 7.5 Spectra for average monthly air temperature readings at Recife, (a) for the raw data; (b) for the seasonally adjusted data using the Tukey window with $M = 24$; (c) the periodogram of the seasonally adjusted data is shown for comparison.

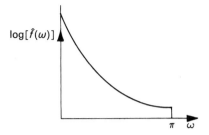

Figure 7.6   The typical spectral shape of an economic time series.

non-parametric procedure in which a finite set of observations is used to estimate a function defined over the range $(0, \pi)$. The function is not constrained to any particular parametric class and so spectral analysis is a more general procedure than inference based on a particular parametric class of models, but is also likely to be less accurate if a parametric model really is appropriate.

Spectral analysis is at its most useful for series of the type shown in Figure 7.4, with no obvious trend or 'seasonal' variation. Such series arise mostly in the physical sciences. In economics, spectral techniques have perhaps not proved as useful as was first hoped, although there have been some successes. Attempts have also been made to apply spectral analysis to marketing data, but it can be argued (Chatfield, 1974) that marketing series are usually too short and the seasonal variation too large for spectral analysis to give useful results. In meteorology and oceanography, spectral analysis can be very useful (e.g. Craddock, 1965; Snodgrass *et al.*, 1966, Mooers and Smith, 1968) but, even in these sciences, spectral analysis may produce no worthwhile results. For example, Chatfield and Pepper (1971) analysed a number of monthly geophysical series but found no tendency to oscillate at frequencies other than the obvious annual effect.

We conclude this section by commenting on some practical aspects of spectral analysis.

Most aspects, such as the choice of truncation point, have already been discussed and will be further clarified in Example 7.1. One problem, which has not been discussed, is whether to plot the estimated spectrum or its logarithm. An advantage of plotting the estimated spectrum on a logarithmic scale is that its asymptotic variance is then independent of the level of the spectrum, so that confidence intervals for the spectrum are of constant width on a logarithmic scale. For spectra showing large variations in power, a logarithmic scale also makes it possible to show more detail over a wide range. (For example, in measuring sound, engineers use decibels which are measured on a logarithmic scale.) Jenkins and Watts (1968, p. 266) suggest that spectrum estimates should always be plotted on a logarithmic scale. But Anderson (1971, p. 547) points out that this exaggerates the visual effects of variations where the spectrum is small. Thus it is often easier to interpret a spectrum plotted on an arithmetic scale as the area under the graph corresponds to power and one can more readily assess the importance of different peaks. So, while it is often useful to plot $\hat{f}(\omega)$ on a logarithmic scale in the initial stages of a spectral analysis, when trying different truncation points and testing the significance of peaks this writer generally prefers to plot the estimated spectrum on an arithmetic scale in order to interpret the final result. It is also generally easier to interpret a spectrum if the frequency scale is measured in cycles per unit time (f) rather than radians per unit time (ω). This has been done in Figures 7.4 and 7.5. A linear transformation of frequency does not affect the **relative** heights of the spectrum at different frequencies, which are of prime importance, though it does change the absolute heights by a constant multiple.

Another point worth mentioning is the possible presence in estimated spectra of **harmonics**. When a spectrum has a large peak at some frequency ω, then related peaks may occur at $2\omega, 3\omega, \ldots$. These multiples of the fundamental frequency are called harmonics and generally speaking simply indicate the non-sinusoidal character of the main cyclical component. For example Mackay (1973) studied the incidence of trips to supermarkets by consumers and found (not surprisingly!) a basic weekly pattern with harmonics at two and three cycles per week.

Finally, a question that is often asked is how large a value of N is required to get a reasonable estimate of the spectrum. It is often recommended that between 100 and 200 observations is the minimum. Granger and Hughes (1968) have tried smaller values of N and conclude that only very large peaks can then be found. If the data are prewhitened, so that the spectrum is fairly flat, then reasonable estimates may be obtained even with values of N less than 100.

Example 7.1   As an example, we analysed part of trace D of Figure 7.4. Although a fairly long trace was available, we decided just to analyse a section lasting for one second to illustrate the problems of analysing a fairly short series. This set of data will also illustrate the problems of analysing a continuous trace as opposed to a discrete series.

The first problem was to digitize the data, and this required the choice of a suitable sampling interval. Inspection of the original trace showed that variation seemed to be 'fairly smooth' over a length of 1 mm, corresponding to $1/100$ second, but to ensure that there was no aliasing we chose $1/200$ second as the sampling interval, giving $N = 200$ observations.

For such a short series, there is little to be gained by using the fast Fourier transform. We decided to transform the truncated autocovariance function, using equation (7.21), with the Tukey window. Several truncation points were tried, and the results for $M = 20$, 40 and 80 are shown in Figure 7.7. Above about 70 cycles per second, the estimates produced by the three values of M are very close to one another when the spectrum is plotted on an arithmetic scale and cannot be distinguished on the graph. Equation (7.21) was evaluated at 51 points, taking $Q=50$, at $\omega=\pi j/50$ ($j=0, 1, \ldots, 50$) where ω is measured in radians per unit time. Now in this example 'unit time' is $1/200$ second and so the values of ω in radians per second are $\omega=200\pi j/50$ for $j=0, 1, \ldots, 50$. Thus the frequencies expressed in cycles per second, by $f=\omega/2\pi$, are $f=2j$ for $j=0, 1, \ldots, 50$. The Nyquist frequency is given by $f_N=100$ cycles per second, which completes one cycle every two observations.

Looking at Figure 7.7, the estimated spectrum is judged rather too smooth with $M=20$, and much too erratic when $M=80$. The value $M=40$ looks about right, although $M=30$ might be even better. There is a clear peak at about 15 cycles per second (15 Hz) as there is in spectrum L of Figure 7.4, but there is also a smaller unexpected peak around 30 cycles per second, which looks like a harmonic of the variation at 15 cycles per second. If a longer series of observations were to be analysed, this peak might disappear, indicating that the peak is spurious.

We also estimated the spectrum using a Parzen window with a truncation point of $M=56$. This value was chosen so that the degrees of freedom of the window, namely 13.3, were almost the same as for the Tukey window with $M=40$. I intended to plot the results on Figure 7.7, but the graph was so close to that produced by the Tukey window that it was impossible to draw them both on the same graph. The biggest difference in estimates at the same frequency was 0.33 at 12 cycles per second, but most of the estimates differed only in the second decimal place. It is clear that the Tukey and Parzen windows give very much the same estimates when equivalent values of M are used.

One feature to note in Figure 7.7 is that the estimated spectrum approaches zero as the frequency approaches the Nyquist frequency. This suggests that there is no aliasing and that our choice of sampling interval was sufficiently small.

Also note that the bandwidths are indicated in Figure 7.7. The bandwidth for the Tukey window is $8\pi/3M$ in radians per unit time. As 'unit time' is $1/200$ second, the bandwidth is $1600\pi/3M$ in radians per second or $800/3M$ in cycles per second.

Confidence intervals can be calculated as described in Section 7.5. For a sample of only 200 observations, they turn out to be disturbingly wide. For example when $M=40$, the degrees of freedom are $2.67N/M=13.3$. For convenience this is rounded off to the nearest integer, namely $v=13$. The peak in the estimated spectrum is at 14 cycles per second, where $\hat{f}(\omega)=7.5$. Here the 95% confidence interval is (3.9 to 19.5). Clearly a longer series is desirable to make the confidence intervals acceptably narrow.

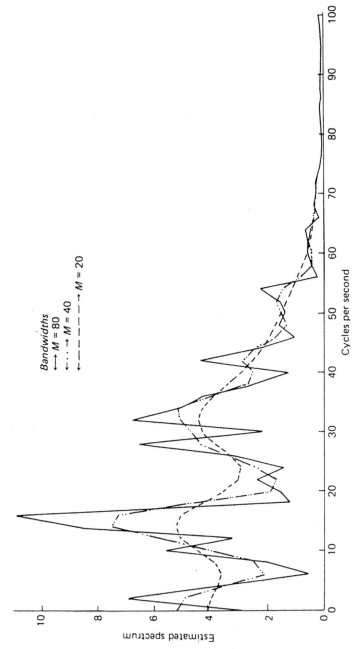

Figure 7.7 Estimated spectra for graph D of Figure 7.4 using the Tukey window with, (a) $M = 80$; (b) $M = 40$; (c) $M = 20$.

EXERCISES

7.1 Revision of Fourier series. Show that the Fourier series which represents the function

$$f(x) = x^2 \qquad \text{in } -\pi \leqslant x \leqslant \pi$$

is given by

$$f(x) = \frac{\pi^2}{3} - 4\left(\frac{\cos x}{1} - \frac{\cos 2x}{2^2} + \frac{\cos 3x}{3^2} - \cdots\right)$$

7.2 Derive equations (7.6) and (7.8).

7.3 Derive Parseval's theorem, given by equation (7.15).

7.4 If $X_1, \ldots, X_N$ are independent $N(\mu, \sigma^2)$ variates show that

$$a_p = 2[\Sigma X_t \cos(2\pi pt/N)]/N$$

is $N(0, 2\sigma^2/N)$ for $p = 1, 2, \ldots, (N/2) - 1$.

7.5 Derive the lag window for smoothing the periodogram in sets of size m. For algebraic simplicity take m odd, with $m = 2m* + 1$, so that

$$\hat{f}(\omega_p) = \frac{1}{m} \sum_{j=-m*}^{m*} I\left(\omega_p + \frac{2\pi j}{N}\right)$$

(Hint: The answer is given in Section 7.4.4. The algebra is messy. Use equation (7.18) and

$$\cos 2\pi k(p+j)/N + \cos 2\pi k(p-j)/N = 2 \cos 2\pi kp/N \cos 2\pi kj/N$$

Also use $\sin A - \sin B = 2 \sin[(A-B)/2]\cos[(A+B)/2]$.)

7.6 Evaluate the degrees of freedom for the Tukey window using equation (7.26).

8

Bivariate processes

Thus far, we have been concerned with analysing a single time series. We now turn our attention to the situation where we have observations on **two** time series and we are interested in the relationship between them.

Jenkins and Watts (1968, p. 322) distinguish two types of situation. In the first type of situation, the two series arise 'on an equal footing' and we are interested in the correlation between them. For example it is often of interest to analyse seismic signals received at two recording sites. In the second, more important type of situation, the two series are 'causally related'. One series is regarded as the **input** to a **linear system**, while the other series is regarded as the **output**; we are then interested in finding the properties of the linear system. The two types of situation are roughly speaking the time-series analogues of **correlation** and **regression**.

The first type of situation is considered in this chapter, where the cross-correlation function and the cross-spectrum are introduced. These functions are also useful in the study of linear systems which are discussed in Chapter 9.

8.1 CROSS-COVARIANCE AND CROSS-CORRELATION FUNCTIONS

Suppose we have N observations on two variables, x and y, at unit time intervals over the same period. The observations will be denoted by $(x_1, y_1), \ldots, (x_N, y_N)$. These observations may be regarded as a finite realization of a discrete bivariate stochastic process (X_t, Y_t).

In order to describe a bivariate process it is useful to know the moments up to second order. For a univariate process, the moments up to second order are the mean and autocovariance function. For a bivariate process, the moments up to second order consist of the mean and autocovariance functions for each of the two components plus a new function, called the **cross-covariance** function, which is given by

$$\gamma_{xy}(t, k) = \text{Cov}(X_t, Y_{t+k})$$

We will only consider bivariate processes which are second-order stationary, so that all moments up to second order do not change with time. We will use the following notation:

$$E(X_t)=\mu_x \qquad E(Y_t)=\mu_y$$
$$\text{Cov}(X_t, X_{t+k})=\gamma_{xx}(k)$$
$$\text{Cov}(Y_t, Y_{t+k})=\gamma_{yy}(k)$$
$$\text{Cov}(X_t, Y_{t+k})=\gamma_{xy}(k) \qquad (8.1)$$

Note that some authors define the cross-covariance function in the 'opposite direction' by

$$\text{Cov}(X_t, Y_{t-k})=\gamma_{xy}^*(k)$$

Comparing with (8.1) we see that

$$\gamma_{xy}(k)=\gamma_{xy}^*(-k)$$

The cross-covariance function differs from the autocovariance function in that it is **not** an even function, since in general

$$\gamma_{xy}(k)\neq\gamma_{xy}(-k)$$

Instead we have the relationship

$$\gamma_{xy}(k)=\gamma_{yx}(-k)$$

where the subscripts are reversed.

   The size of the cross-covariance coefficients depends on the units in which X_t and Y_t are measured. Thus for interpretative purposes, it is useful to standardize the cross-covariance function to produce a function called the **cross-correlation** function, $\rho_{xy}(k)$, which is defined by

$$\rho_{xy}(k)=\gamma_{xy}(k)/\sqrt{[\gamma_{xx}(0)\gamma_{yy}(0)]} \qquad (8.2)$$

This function measures the correlation between X_t and Y_{t+k} and has the properties

(a) $\rho_{xy}(k)=\rho_{yx}(-k)$

(b) $|\rho_{xy}(k)|\leqslant 1$ (see Exercise 8.2).

Whereas $\rho_{xx}(0)$, $\rho_{yy}(0)$ are both equal to one, the value of $\rho_{xy}(0)$ is **not** necessarily equal to one, a fact which is sometimes overlooked.

8.1.1 Examples

Before discussing the estimation of cross-covariance and cross-correlation functions, we will derive the theoretical functions for two examples of bivariate processes. The first example is rather 'artificial', but the model in Example 8.2 can be useful in practice.

Example 8.1   Suppose that $\{X_t\}, \{Y_t\}$ are both formed from the same purely random process $\{Z_t\}$, which has mean zero, variance σ_Z^2, by

$$X_t = Z_t$$
$$Y_t = 0.5Z_{t-1} + 0.5Z_{t-2}$$

Then using (8.1) we have

$$\gamma_{xy}(k) = \begin{cases} 0.5\sigma_Z^2 & k = 1, 2 \\ 0 & \text{otherwise} \end{cases}$$

Now the variances of the two components are given by

$$\gamma_{xx}(0) = \sigma_Z^2$$
$$\gamma_{yy}(0) = \sigma_Z^2/2$$

so that, using (8.2), we have

$$\rho_{xy}(k) = \begin{cases} 0.5\sqrt{2} & k = 1, 2 \\ 0 & \text{otherwise} \end{cases}$$

Example 8.2   Suppose that

$$X_t = Z_{1,t}$$
$$Y_t = X_{t-d} + Z_{2,t} \tag{8.3}$$

where $\{Z_{1,t}\}, \{Z_{2,t}\}$ are uncorrelated purely random processes with mean zero and variance σ_Z^2, and where d is an integer. Then we find

$$\gamma_{xy}(k) = \begin{cases} \sigma_Z^2 & k = d \\ 0 & \text{otherwise} \end{cases}$$

$$\rho_{xy}(k) = \begin{cases} 1/\sqrt{2} & k = d \\ 0 & \text{otherwise} \end{cases}$$

In Chapter 9 we will see that equation (8.3) corresponds to putting noise into a linear system which consists of a simple delay of lag d and then adding more noise. The cross-correlation function has a peak at lag d corresponding to the delay in the system, a result which the reader should find intuitively reasonable.

8.1.2   Estimation

The 'obvious' way of estimating the cross-covariance and cross-correlation functions is by means of corresponding sample functions. With N pairs of observations $\{(x_i, y_i); i = 1 \text{ to } N\}$, the sample cross-covariance function is

$$c_{xy}(k) = \begin{cases} \sum_{t=1}^{N-k} (x_t - \bar{x})(y_{t+k} - \bar{y})/N & k = 0, 1, \ldots, N-1 \\ \sum_{t=1-k}^{N} (x_t - \bar{x})(y_{t+k} - \bar{y})/N & \begin{aligned} k = & -1, -2, \ldots, \\ & -(N-1) \end{aligned} \end{cases}$$ (8.4)

and the sample cross-correlation function is

$$r_{xy}(k) = c_{xy}(k)/\sqrt{[c_{xx}(0)c_{yy}(0)]}$$ (8.5)

where $c_{xx}(0)$, $c_{yy}(0)$ are the sample variances of observations on x_t and y_t respectively.

It can be shown that these estimators are asymptotically unbiased and consistent. However, it can also be shown that successive estimates are themselves autocorrelated. In addition the variances of the estimators depend on the autocorrelation functions of the two components. Thus for moderately large values of N (e.g. N about 100) it is possible for two series, which are actually unrelated, to give rise to apparently 'large' cross-correlation coefficients which are actually spurious. Thus if a test is required for non-zero correlation between two time series, both series should first be filtered to convert them to white noise before computing the cross-correlation function (Jenkins and Watts, 1968, p. 340). For two uncorrelated series of white noise it can be shown that

$$E[r_{xy}(k)] \simeq 0$$

$$\mathrm{Var}[r_{xy}(k)] \simeq 1/N$$

so that values outside the interval $\pm 2/\sqrt{N}$ are significantly different from zero.

The filtering procedure mentioned above is accomplished by treating each series separately and fitting an appropriate model. The new filtered series consists of the residuals from this model. For example, suppose that one series appeared to be a first-order AR process with estimated parameter $\hat{\alpha}$. Then the filtered series is given by

$$x'_t = (x_t - \bar{x}) - \hat{\alpha}(x_{t-1} - \bar{x})$$

8.1.3 Interpretation

The interpretation of the sample cross-correlation function can be fraught with danger unless one uses the prefiltering procedure described in Section 8.1.2. For example, Coen, Gomme and Kendall (1969) calculated cross-correlation functions between variables such as (detrended) *Financial Times* (*FT*) share index and (detrended) UK car production, and this resulted in a fairly smooth, roughly sinusoidal function with 'large' coefficients at lags 5 and 6. Coen *et al.* used this information to set up a regression model to 'explain' the variation in the *FT* share index. However, Box and Newbold (1971) have shown that the 'large' cross-correlation coefficients are spurious as the two series had not been properly filtered. By simply detrending the raw data, Coen *et al.* effectively assumed that the appropriate model for each series was of the form

$$x_t = \alpha + \beta t + a_t$$

where the a_t are **independent**. In fact the error structure was quite different.

   If both series are properly filtered, we have seen that it is easy to test whether any of the cross-correlation coefficients are significantly different from zero. Following Example 8.2, a peak in the estimated cross-correlation function at lag d may indicate that one series is related to the other when delayed by time d.

8.2   THE CROSS-SPECTRUM

The cross-correlation function is the natural tool for examining the relationship between two time series in the time domain. In this section we introduce a complementary function, called the cross spectral density function or cross-spectrum, which is the natural tool in the frequency domain.

   By analogy with equation (6.11), we will define the cross-spectrum of a discrete bivariate process measured at unit intervals of time as the Fourier transform of the cross-covariance function, namely

$$f_{xy}(\omega) = \frac{1}{\pi}\left[\sum_{k=-\infty}^{\infty} \gamma_{xy}(k)e^{-i\omega k} \right] \tag{8.6}$$

over the range $0 < \omega < \pi$. The physical interpretation of the cross-spectrum is more difficult than for the autospectrum (see Priestley, 1981, p. 657). Indeed a physical understanding of cross-spectra will probably not become clear until we have studied linear systems.

   Note that $f_{xy}(\omega)$ is a **complex** function, unlike the autospectrum which is real. This is because $\gamma_{xy}(k)$ is not an even function.

   The reader should note that many authors define the cross-spectrum in the range $(-\pi, \pi)$ by analogy with equation (6.13) as

$$f_{xy}(\omega) = \frac{1}{2\pi}\left[\sum_{k=-\infty}^{\infty} \gamma_{xy}(k)e^{-i\omega k} \right] \tag{8.7}$$

This definition has certain mathematical advantages, notably that it can handle complex-valued processes and that it has a simple inverse relationship of the form

$$\gamma_{xy}(k) = \int_{-\pi}^{\pi} e^{i\omega k} f_{xy}(\omega)\, d\omega \tag{8.8}$$

whereas (8.6) does not have a simple inverse relationship. But (8.7) introduces negative frequencies, and for ease of understanding we shall use (8.6). As regards definition (8.7), note that $f_{xy}(-\omega)$ is the complex conjugate of $f_{xy}(\omega)$ and so provides no extra information. Authors who use (8.7) only examine $f_{xy}(\omega)$ at positive frequencies. Note that Kendall, Stuart and Ord (1983, Chapter 51) use a different definition to other authors by omitting the constant $1/\pi$ and transforming the cross-correlation function rather than the cross-covariance function.

We now describe several functions derived from the cross-spectrum which are helpful in interpreting the cross-spectrum. From (8.6), the real part of the cross-spectrum, called the **co-spectrum**, is given by

$$c(\omega) = \frac{1}{\pi} \left[\sum_{k=-\infty}^{\infty} \gamma_{xy}(k)\cos \omega k \right]$$

$$= \frac{1}{\pi} \left\{ \gamma_{xy}(0) + \sum_{k=1}^{\infty} \left[\gamma_{xy}(k) + \gamma_{yx}(k) \right] \cos \omega k \right\} \tag{8.9}$$

The complex part of the cross-spectrum, with a minus sign, is called the **quadrature** spectrum and is given by

$$q(\omega) = \frac{1}{\pi} \left[\sum_{k=-\infty}^{\infty} \gamma_{xy}(k)\sin \omega k \right]$$

$$= \frac{1}{\pi} \left\{ \sum_{k=1}^{\infty} \left[\gamma_{xy}(k) - \gamma_{yx}(k) \right] \sin \omega k \right\} \tag{8.10}$$

so that

$$f_{xy}(\omega) = c(\omega) - iq(\omega) \tag{8.11}$$

Note that Kendall, Stuart and Ord (1983, Section 51.33) express the cross-spectrum as $c(\omega) + iq(\omega)$ in view of their alternative definition of the cross-covariance function (as $\gamma_{xy}^*(k)$ — see Section 8.1).

An alternative way of expressing the cross-spectrum is in the form

$$f_{xy}(\omega) = \alpha_{xy}(\omega) e^{i\phi_{xy}(\omega)} \tag{8.12}$$

where

$$\alpha_{xy}(\omega) = \sqrt{[c^2(\omega) + q^2(\omega)]} \tag{8.13}$$

is the cross-amplitude spectrum, and

$$\phi_{xy}(\omega) = \tan^{-1}[-q(\omega)/c(\omega)] \tag{8.14}$$

is the **phase** spectrum. From (8.14) it appears that $\phi_{xy}(\omega)$ is undetermined by a multiple of π. However, if the cross-amplitude spectrum is required to be positive so that we take the positive square root in (8.13), then it can be seen that the phase is actually undetermined by a multiple of 2π using the equality of (8.11) and (8.12). This apparent non-uniqueness makes it difficult to graph the phase. However, when we consider linear systems in Chapter 9, we will see that there are physical reasons why the phase **is** generally uniquely determined and not confined to the range $\pm\pi$ or $\pm\pi/2$. The phase is usually zero at $\omega=0$ and is a continuous function as ω goes from 0 to π.

Another useful function derived from the cross-spectrum is the (squared) **coherency**, which is given by

$$C(\omega) = [c^2(\omega) + q^2(\omega)]/[f_x(\omega)f_y(\omega)]$$
$$= \alpha_{xy}^2(\omega)/f_x(\omega)f_y(\omega) \qquad (8.15)$$

where $f_x(\omega), f_y(\omega)$ are the power spectra of the individual processes, X_t and Y_t. It can be shown that

$$0 \leqslant C(\omega) \leqslant 1$$

This quantity measures the square of the linear correlation between the two components of the bivariate process at frequency ω and is analogous to the square of the usual correlation coefficient. The closer $C(\omega)$ is to one, the more closely related are the two processes at frequency ω.

Finally we will define a function called the **gain** spectrum, which is given by

$$G_{xy}(\omega) = \sqrt{[f_y(\omega)C(\omega)/f_x(\omega)]}$$
$$= \alpha_{xy}(\omega)/f_x(\omega) \qquad (8.16)$$

which is essentially the regression coefficient of the process Y_t on the process X_t at frequency ω. A second gain function can also be defined by $G_{yx}(\omega) = \alpha_{xy}(\omega)/f_y(\omega)$ in which, using linear system terminology, Y_t is regarded as the input and X_t as the output.

By this point, the reader will probably be rather confused by all the different functions which have been introduced in relation to the cross-spectrum. Whereas the cross-correlation function is a relatively straightforward development from the autocorrelation function, statisticians often find the cross-spectrum much harder to understand than the autospectrum. Usually **three** functions have to be plotted against frequency to describe the relationship between two series in the frequency domain. Sometimes the co-, quadrature and coherency spectra are most suitable. Sometimes the coherency, phase and cross-amplitude are more appropriate, while another possible trio is coherency, phase and gain. The physical interpretation of these functions will probably not become clear until we have studied linear systems in Chapter 9.

8.2.1   Examples

In this section, we derive the cross-spectrum and related functions for the two examples discussed in Section 8.1.1.

Example 8.3   For Example 8.1, using (8.6) the cross-spectrum is given by

$$f_{xy}(\omega) = (0.5\sigma_Z^2 \, e^{-i\omega} + 0.5\sigma_Z^2 \, e^{-2i\omega})/\pi$$

Using (8.9), the co-spectrum is given by

$$c(\omega) = 0.5\sigma_Z^2(\cos \omega + \cos 2\omega)/\pi$$

Using (8.10), the quadrature spectrum is given by

$$q(\omega) = 0.5\sigma_Z^2(\sin \omega + \sin 2\omega)/\pi$$

Using (8.13), the cross-amplitude spectrum is given by

$$\alpha_{xy}(\omega) = \frac{0.5\sigma_Z^2}{\pi} \sqrt{[(\cos \omega + \cos 2\omega)^2 + (\sin \omega + \sin 2\omega)^2]}$$

which, after some algebra, gives

$$\alpha_{xy}(\omega) = \sigma_Z^2 \cos(\omega/2)/\pi$$

Using (8.14), the phase spectrum is given by

$$\tan \phi_{xy}(\omega) = -(\sin \omega + \sin 2\omega)/(\cos \omega + \cos 2\omega)$$

In order to evaluate the coherency, we need to find the power spectra of the two processes. Since $X_t = Z_t$, it has a constant spectrum given by $f_x(\omega) = \sigma_Z^2/\pi$. The spectrum of Y_t is given by

$$f_y(\omega) = 0.5\sigma_Z^2(1 + \cos \omega)/\pi$$
$$= \sigma_Z^2 \cos^2(\omega/2)/\pi$$

Thus, using (8.15), the coherency spectrum is given by

$$C(\omega) = 1 \qquad \text{for all } \omega \text{ in } (0, \pi)$$

This latter result may at first sight appear surprising. But both X_t and Y_t are generated from the **same** noise process and this explains why there is perfect correlation between the components of the two processes at any given frequency. Finally, using (8.16), the gain spectrum is given by

$$G_{xy}(\omega) = \cos(\omega/2)$$

Since the coherency is unity, $G_{xy}(\omega) = \sqrt{[f_y(\omega)/f_x(\omega)]}$.

Example 8.4   For Example 8.2 we find

$$f_{xy}(\omega) = \sigma_Z^2 \, e^{-i\omega d}/\pi$$

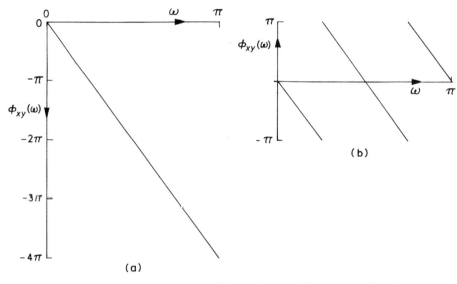

Figure 8.1 The phase spectrum for Example 8.4 with $d=4$, with, (a) phase unconstrained; (b) phase constrained.

$$c(\omega) = \sigma_Z^2 \cos \omega d/\pi$$

$$q(\omega) = \sigma_Z^2 \sin \omega d/\pi$$

$$\alpha_{xy}(\omega) = \sigma_Z^2/\pi$$

$$\tan \phi_{xy}(\omega) = -\tan \omega d \qquad (8.17)$$

Then, as the two autospectra are given by

$$f_x(\omega) = \sigma_Z^2/\pi$$

$$f_y(\omega) = 2\sigma_Z^2/\pi$$

we find

$$C(\omega) = 1/2$$

The function of particular interest in this example is the phase, which from (8.17) is a straight line with slope $-d$ when $\phi_{xy}(\omega)$ is unconstrained and is plotted against ω as a continuous function starting with zero phase at zero frequency (see Figure 8.1(a)). If, however, the phase is constrained to lie within the interval $(-\pi, \pi)$ then a graph like Figure 8.1(b) will result, where the slope of each line is $-d$.

This result is often used in identifying relationships between time series. If the estimated phase approximates a straight line through the origin then this indicates a delay between the two series equal to the slope of the line. An example of this in practice is given by Barksdale and Guffey (1972). More generally, the time delay between two recording sites will change with frequency, due, for example, to varying speeds of propagation. This is called the **dispersive** case, and real-life examples are discussed by Haubrich and Mackenzie (1965) and Hamon and Hannan (1974).

8.2.2 Estimation

As in Section 7.4, there are two basic approaches to estimating the cross-spectrum. First, we can take a Fourier transform of the truncated sample cross-covariance function (or of the cross-correlation function to get a normalized cross-spectrum). The estimated co-spectrum is then given by

$$\hat{c}(\omega) = \frac{1}{\pi} \left[\sum_{k=-M}^{M} \lambda_k c_{xy}(k) \cos \omega k \right] \tag{8.18}$$

where M is the truncation point, and $\{\lambda_k\}$ is the lag window. The estimated quadrature spectrum is given by

$$\hat{q}(\omega) = \frac{1}{\pi} \left[\sum_{k=-M}^{M} \lambda_k c_{xy}(k) \sin \omega k \right] \tag{8.19}$$

Equations (8.18) and (8.19) are often used in the form

$$\hat{c}(\omega) = \frac{1}{\pi} \left\{ \lambda_0 c_{xy}(0) + \sum_{k=1}^{M} \lambda_k [c_{xy}(k) + c_{xy}(-k)] \cos \omega k \right\}$$

$$\hat{q}(\omega) = \frac{1}{\pi} \left\{ \sum_{k=1}^{M} \lambda_k [c_{xy}(k) - c_{xy}(-k)] \sin \omega k \right\}$$

The truncation point M and the lag window $\{\lambda_k\}$ are chosen in a similar way to that used in spectral analysis for a single series, with the Tukey and Parzen windows being most popular.

Having estimated the co- and quadrature spectra, estimates of the cross-amplitude spectrum, phase and coherency follow in an obvious way from equations (8.13), (8.14) and (8.15). We have

$$\hat{\alpha}_{xy}(\omega) = \sqrt{[\hat{c}^2(\omega) + \hat{q}^2(\omega)]}$$

$$\tan \hat{\phi}_{xy}(\omega) = -\hat{q}(\omega)/\hat{c}(\omega)$$

$$\hat{C}(\omega) = \hat{\alpha}_{xy}^2(\omega)/\hat{f}_x(\omega)\hat{f}_y(\omega)$$

When plotting the estimated phase spectrum, similar remarks apply as to the (theoretical) phase. Phase estimates are apparently not uniquely determined but can usually be plotted as a continuous function which is zero at zero frequency.

Before estimating the coherency, it may be advisable to **align** the two series. If this is not done, Jenkins and Watts (1968) have demonstrated that estimates of coherency will be biased if the phase changes rapidly. If the sample cross-correlation function has its largest value at lag s, then the two series are aligned by translating one of the series a distance s so that the peak in the cross-correlation function of the aligned series is at zero lag.

The second approach to cross-spectral analysis is to smooth a function called the **cross-periodogram**. The univariate periodogram of a series $\{x_t\}$ can be written in the form

$$I(\omega_p) = (\Sigma x_t\, e^{i\omega_p t})\, (\Sigma x_t\, e^{-i\omega_p t})/N\pi$$
$$= N(a_p^2 + b_p^2)/4\pi \tag{8.20}$$

using (7.17) and (7.10). By analogy with (8.20) we may define the cross-periodogram of two series (x_t) and (y_t) as

$$I_{xy}(\omega_p) = (\Sigma x_t\, e^{i\omega_p t})\, (\Sigma y_t\, e^{-i\omega_p t})/N\pi \tag{8.21}$$

We then find that the real and imaginary parts of $I_{xy}(\omega_p)$ are given by

$$N(a_{px}a_{py} + b_{px}b_{py})/4\pi \quad \text{and} \quad N(a_{px}b_{py} - a_{py}b_{px})/4\pi$$

where (a_{px}, b_{px}), (a_{py}, b_{py}) are the Fourier coefficients of $\{x_t\}$, $\{y_t\}$ at ω_p. These real and imaginary parts may then be smoothed to get consistent estimates of the co- and quadrature spectral density functions by

$$\hat{c}(\omega_p) = N \sum_{q=p-m^*}^{p+m^*} (a_{qx}a_{qy} + b_{qx}b_{qy})/4\pi m$$

$$\hat{q}(\omega_p) = N \sum_{q=p-m^*}^{p+m^*} (a_{qx}b_{qy} - a_{qy}b_{qx})/4\pi m$$

where $m = 2m^* + 1$. These estimates may then be used to estimate the cross-amplitude spectrum, phase etc. as before.

The computational advantages of this type of approach are clear. Once a periodogram analysis has been made of the two individual processes, nearly all the work has been done as the estimates of $c(\omega)$ and $q(\omega)$ only involve the Fourier coefficients of the two series. The disadvantage of the approach is that alignment is only possible if the cross-correlation function is calculated separately. This can be done directly or by the use of two (fast) Fourier transforms by an analogous procedure to that described in Section 7.4.5.

The properties of cross-spectral estimators are discussed by Jenkins and Watts (1968) and Priestley (1981), while Granger and Hughes (1968) have carried out a simulation study on some short series. The following points are worth noting. Estimates of phase and cross-amplitude are imprecise when the coherency is relatively small. Estimates of coherency are constrained to lie between 0 and 1, and there may be a bias towards 1/2 which may be serious with short series. Finally we note that rapid changes in phase may bias coherency estimates.

8.2.3   Interpretation

Cross-spectral analysis is a technique for examining the relationship between two series over a range of frequencies. The technique may be used for two time series which 'arise on a similar footing' and then the coherency spectrum is perhaps the most useful function. It measures the linear correlation between two series at each frequency and is analogous to the square of the ordinary product-moment correlation coefficient.

The other functions introduced in this chapter, such as the phase spectrum, are most readily understood in the context of linear systems which will be discussed in Chapter 9. We will therefore defer further discussion of how to interpret cross-spectral estimates until Section 9.3.

EXERCISES

8.1   Show that the cross-covariance function of the discrete bivariate process (X_t, Y_t) where

$$X_t = Z_{1,t} + \beta_{11}Z_{1,t-1} + \beta_{12}Z_{2,t-1}$$
$$Y_t = Z_{2,t} + \beta_{21}Z_{1,t-1} + \beta_{22}Z_{2,t-1}$$

and $(Z_{1,t})$, $(Z_{2,t})$ are independent purely random processes with zero mean and variance σ_Z^2, is given by

$$\gamma_{xy}(k) = \begin{cases} \sigma_Z^2(\beta_{11}\beta_{21} + \beta_{12}\beta_{22}) & k=0 \\ \beta_{21}\sigma_Z^2 & k=1 \\ \beta_{12}\sigma_Z^2 & k=-1 \\ 0 & \text{otherwise} \end{cases}$$

Hence evaluate the cross-spectrum.

8.2   Define the cross-correlation function $\rho_{xy}(\tau)$ of a bivariate stationary process and show that $|\rho_{xy}(\tau)| \leqslant 1$ for all τ. Two MA processes

$$X_t = Z_t + 0.4Z_{t-1}$$
$$Y_t = Z_t - 0.4Z_{t-1}$$

are formed from a purely random process $\{Z_t\}$ which has mean zero and variance σ_Z^2. Find the cross-covariance and cross-correlation functions of the bivariate process $\{X_t,\ Y_t\}$ and hence show that the cross-spectrum is given by

$$f_{xy}(\omega) = \sigma_Z^2(0.84 + 0.8i \sin \omega)/\pi \qquad 0 < \omega < \pi$$

Evaluate the co-, quadrature, cross-amplitude, phase and coherency spectra.

9

Linear systems

9.1 INTRODUCTION

An important problem in engineering and the physical sciences is that of identifying a model for a physical system (or process) given observations on the input and output to the system. For example, the yield from a chemical reactor (the output) depends *inter alia* on the temperature at which the reactor is kept (the input). Much of the literature assumes that the system can be adequately approximated over the range of interest by a linear model whose parameters do not change with time, although recently there has been increased interest in time-varying and non-linear systems. Some general references include Eykhoff (1974), Harris (1976) and Ljung (1987).

We shall denote the input and output series by $\{x_t\}$, $\{y_t\}$ respectively in discrete time, and by $\{x(t)\}$, $\{y(t)\}$ respectively in continuous time.

In this chapter we confine attention to linear systems. A precise definition of **linearity** is as follows. Suppose $y_1(t)$, $y_2(t)$ are the outputs corresponding to $x_1(t)$, $x_2(t)$ respectively. Then the system is said to be linear if, and only if, a linear combination of the inputs, say $\lambda_1 x_1(t) + \lambda_2 x_2(t)$, produces the same linear combination of the outputs, namely $\lambda_1 y_1(t) + \lambda_2 y_2(t)$, where λ_1, λ_2 are any constants.

We shall further confine our attention to linear systems which are **time-invariant**. This term is defined as follows. If input $x(t)$ produces output $y(t)$, then the system is said to be time-invariant if a delay of time τ in the input produces the same delay in the output. In other words $x(t-\tau)$ produces output $y(t-\tau)$, so that the input–output relation does not change with time.

We will only consider systems having one input and one output. The extension to several inputs and outputs is straightforward in principle, though difficult in practice.

The study of linear systems is useful not only for the purpose of examining the relationship between different time series, but also for examining the properties of filtering procedures such as detrending.

In Sections 9.2 and 9.3 we show how to describe linear systems in the time and frequency domains respectively, while in Section 9.4 we discuss the identification of linear systems from observed data.

9.2   LINEAR SYSTEMS IN THE TIME DOMAIN

A time-invariant linear system may generally be written in the form

$$y(t) = \int_{-\infty}^{\infty} h(u)x(t-u)\, du \qquad (9.1)$$

in continuous time, or

$$y_t = \sum_{k=-\infty}^{\infty} h_k x_{t-k} \qquad (9.2)$$

in discrete time. The weight function, $h(u)$ in continuous time or $\{h_k\}$ in discrete time, provides a description of the system in the time domain. This function is called the **impulse response function** of the system, for reasons which will become apparent later.

It is clear that equations (9.1) and (9.2) are linear. The property of time invariance ensures that the impulse response function does not depend on t. The system is said to be **physically realizable** or causal if

$$h(u) = 0 \qquad u < 0$$

or

$$h_k = 0 \qquad k < 0$$

Engineers have been principally concerned with continuous-time systems but are increasingly studying sampled-data control problems. Statisticians generally work with discrete data and so the subsequent discussion is mainly concerned with the discrete case.

We will only consider **stable** systems for which any bounded input produces a bounded output, although control engineers are frequently concerned with the control of unstable systems. A sufficient condition for stability is that the impulse response function should satisfy

$$\sum_k |h_k| < C$$

where C is a finite constant.

9.2.1   Some types of linear system

The linear filters introduced in Section 2.5.2 are examples of linear systems. For example the simple moving average given by

$$y_t = (x_{t-1} + x_t + x_{t+1})/3$$

has impulse response function

$$h_k = \begin{cases} 1/3 & k = -1, 0, +1 \\ 0 & \text{otherwise} \end{cases}$$

Note that this filter is not 'physically realizable', although it can of course be used as a mathematical smoothing device.

Another general class of linear systems are those expressed as linear differential equations with constant coefficients in continuous time. For example, if T is a constant, then

$$T\frac{dy(t)}{dt} + y(t) = x(t)$$

is a description of a linear system. In discrete time, the analogue of differential equations are **difference** equations given by

$$y_t + \alpha_1 \nabla y_t + \alpha_2 \nabla^2 y_t + \cdots = \beta_0 x_t + \beta_1 \nabla x_t + \beta_2 \nabla^2 x_t + \cdots \qquad (9.3)$$

where $\nabla y_t = y_t - y_{t-1}$. Equation (9.3) can be rewritten as

$$y_t = a_1 y_{t-1} + a_2 y_{t-2} + \cdots + b_0 x_t + b_1 x_{t-1} + \cdots \qquad (9.4)$$

It is clear that equation (9.4) can be rewritten in the form (9.2) by successive substitution. For example if

$$y_t = \tfrac{1}{2} y_{t-1} + x_t$$

then we find

$$y_t = x_t + \tfrac{1}{2} x_{t-1} + \tfrac{1}{4} x_{t-2} \cdots$$

so that the impulse response function is given by

$$h_k = \begin{cases} (\tfrac{1}{2})^k & k=0, 1, \ldots \\ 0 & k < 0 \end{cases}$$

Two very simple linear systems are given by

$$y_t = x_{t-d} \qquad (9.5)$$

called **simple delay**, where the integer d denotes the delay time, and

$$y_t = g x_t \qquad (9.6)$$

called **simple gain**, where g is a constant called the gain. The impulse response functions of (9.5) and (9.6) are

$$h_k = \begin{cases} 1 & k=d \\ 0 & \text{otherwise} \end{cases}$$

and

$$h_k = \begin{cases} g & k=0 \\ 0 & \text{otherwise} \end{cases}$$

respectively.

In continuous time, the impulse response functions of simple delay and simple gain, namely

$$y(t)=x(t-\tau)$$

and

$$y(t)=gx(t)$$

can only be represented in terms of the Dirac delta function (see Appendix B). The functions are $\delta(u-\tau)$ and $g\delta(u)$ respectively.

An important class of impulse response functions, which often provides a reasonable approximation to physically realizable systems, is given by

$$h(u)=\begin{cases}[g\,e^{-(u-\tau)/T}]/T & u>\tau \\ 0 & u<\tau\end{cases}$$

A function of this type is called a **delayed exponential**, and depends on three constants, g, T and τ. The constant τ is called the **delay**. When $\tau=0$, we have simple exponential response. The constant g is called the **gain**, and represents the eventual change in output when a step change of unit size is made to the input. The constant T governs the rate at which the output changes. Figure 9.1 shows how the output to a delayed exponential system changes when a step change of unity is made to the input.

9.2.2   The impulse response function

The impulse response function describes how the output is related to the input of a linear system (see equations (9.1) and (9.2)). The name 'impulse response' arises from the fact that the function describes the response of the system to an impulse input of unit size. For example, in discrete time, suppose that the input x_t is zero for all t except at time zero when it takes the value unity, so that $x_0=1$. Then the output at time t is given by

$$y_t=\Sigma h_k x_{t-k}$$

$$=h_t$$

Thus the output resulting from the unit impulse input is the same as the impulse response function, and this explains why engineers often prefer the description 'unit impulse response function'.

9.2.3   The step response function

An alternative, equivalent, way of describing a linear system in the time domain is by means of a function called the step response function, which is defined by

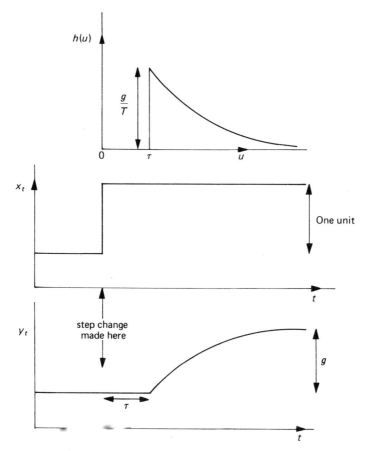

Figure 9.1 A delayed exponential response to a unit step change in input, showing graphs of, (a) impulse response function; (b) input; (c) output.

$$S(t) = \int_{-\infty}^{t} h(u) \, du \qquad (9.7)$$

in continuous time, and

$$S_t = \sum_{k \leqslant t} h_k \qquad (9.8)$$

in discrete time.

The name 'step response' arises from the fact that the function describes the response of the system to a unit step change in the input. For example, in discrete time, suppose that the input is given by

$$x_t = \begin{cases} 0 & t < 0 \\ 1 & t \geqslant 0 \end{cases}$$

Then

$$y_t = \sum_k h_k x_{t-k} = \sum_{k \leqslant t} h_k = S_t$$

so that the output is equal to the step response function.

Engineers sometimes use this relationship to measure the properties of a physically realizable system. The input is held steady for some time and then a unit step change is made to the input. The output is then observed and this provides an estimate of the step response function, and hence of its derivative, the impulse response function. A step change in the input may be easier to provide than an impulse.

The step response function for a delayed exponential system is given by

$$S(t) = g[1 - e^{-(t-\tau)/T}] \qquad t > \tau \tag{9.9}$$

and the graph of $y(t)$ in Figure 9.1 is also a graph of $S(t)$.

9.3   LINEAR SYSTEMS IN THE FREQUENCY DOMAIN

9.3.1   The frequency response function

An alternative way of describing a linear system is by means of a function, called the frequency response function or transfer function, which is the Fourier transform of the impulse response function. It is defined by

$$H(\omega) = \int_{-\infty}^{\infty} h(u)e^{-i\omega u}\, du \qquad 0 < \omega < \infty \tag{9.10}$$

in continuous time, and

$$H(\omega) = \sum_k h_k e^{-i\omega k} \qquad 0 < \omega < \pi \tag{9.11}$$

in discrete time.

The frequency response and impulse response functions are equivalent ways of describing a linear system, in a somewhat similar way that the autocovariance and power spectral density functions are equivalent ways of describing a stationary stochastic process, one function being the Fourier transform of the other. We shall see that, for some purposes, $H(\omega)$ is much more useful than $h(u)$. First we prove the following theorem.

Theorem 9.1   A sinusoidal input to a linear system gives rise, in the steady state, to a sinusoidal output at the **same** frequency. The amplitude of the sinusoid may change and there may also be a phase shift.

Proof   The proof is given for continuous time, the extension to discrete time being straightforward. Suppose that the input to a linear system, with impulse response function $h(u)$, is given by

$$x(t) = \cos \omega t \qquad \text{for all } t$$

Then the output is given by

$$y(t) = \int_{-\infty}^{\infty} h(u) \cos \omega(t - u) \, du \qquad (9.12)$$

Now $\cos(A - B) = \cos A \cos B + \sin A \sin B$, so we may rewrite (9.12) as

$$y(t) = \cos \omega t \int_{-\infty}^{\infty} h(u) \cos \omega u \, du + \sin \omega t \int_{-\infty}^{\infty} h(u) \sin \omega u \, du$$

As the two integrals do not depend on t, it is now obvious that $y(t)$ is a mixture of sine and cosine terms at frequency ω. Thus the output is a sinusoidal perturbation at the same frequency ω as the input.
    If we write

$$A(\omega) = \int_{-\infty}^{\infty} h(u) \cos \omega u \, du$$

$$B(\omega) = \int_{-\infty}^{\infty} h(u) \sin \omega u \, du$$

$$G(\omega) = \sqrt{[A^2(\omega) + B^2(\omega)]} \qquad (9.13)$$

$$\tan \phi(\omega) = -B(\omega)/A(\omega) \qquad (9.14)$$

then

$$y(t) = A(\omega) \cos \omega t + B(\omega) \sin \omega t$$

$$= G(\omega) \cos[\omega t + \phi(\omega)] \qquad (9.15)$$

Equation (9.15) shows that a cosine wave is amplified by a factor $G(\omega)$, which is called the **gain** of the system. The equation also shows that the cosine wave is shifted by an angle $\phi(\omega)$, which is called the **phase shift**. Note that both the gain and phase shift may vary with frequency. From equation (9.14) we see that the phase shift is apparently not uniquely determined. If we take the positive square root in equation (9.13), so that the gain is required to be positive, then the phase shift is undetermined by a multiple of 2π (see also Sections 8.2 and 9.3.2).
    We have so far considered an input cosine wave. By a similar argument it can be shown that an input sine wave, $x(t) = \sin \omega t$, gives an output $y(t) = G(\omega) \sin[\omega t + \phi(\omega)]$, so that there is the same gain and phase shift. More generally if we consider an input given by

$$x(t) = e^{i\omega t} = \cos \omega t + i \sin \omega t$$

then the output is given by

$$y(t) = G(\omega)\{\cos[\omega t + \phi(\omega)] + i \sin[\omega t + \phi(\omega)]\}$$
$$= G(\omega)e^{i[\omega t + \phi(\omega)]}$$
$$= G(\omega)e^{i\phi(\omega)}x(t) \qquad (9.16)$$

Now from equations (9.13) and (9.14)

$$G(\omega)e^{i\phi(\omega)} = A(\omega) - iB(\omega)$$
$$= \int_{-\infty}^{\infty} h(u)(\cos \omega u - i \sin \omega u)\, du$$
$$= \int_{-\infty}^{\infty} h(u)e^{-i\omega u}\, du$$
$$= H(\omega) \qquad (9.17)$$

So when the input in equation (9.16) is of the form $e^{i\omega t}$, the output is given simply by frequency response function times input, and we have (in the steady-state situation)

$$y(t) = H(\omega)x(t) \qquad (9.18)$$

This completes the proof of Theorem 9.1.

Transients   The reader should note that Theorem 9.1 only applies in the **steady state** where it is assumed that the input sinusoid was applied at $t = -\infty$. If in fact the sinusoid is applied at say $t = 0$, then the output will take some time to settle to the steady-state form given by the theorem. The difference between the observed output and the steady-state output is called the **transient** component. The system is stable if this transient component tends to zero as $t \to \infty$. If the relationship between input and output is expressed as a differential (or difference) equation, then the steady-state solution corresponds to the particular integral, while the transient component corresponds to the complementary function.

It is easier to describe the transient behaviour of a linear system by using the **Laplace** transform of the impulse response function. Engineers also prefer the Laplace transform as it is defined for unstable systems. However, statisticians have customarily dealt with steady-state behaviour and used Fourier transforms, and we will continue this custom. Nevertheless this is certainly an aspect of linear systems which statisticians should look at more closely.

Discussion of Theorem 9.1   Theorem 9.1 helps to introduce the importance of the frequency response function. For inputs consisting of an impulse or step change it is easy to calculate the output using the impulse response function. But for a sinusoidal input, it is much easier to calculate the output using the frequency response function. (Compare equations (9.12) and (9.18).) More generally for an input consisting of several sinusoidal perturbations, namely

$$x(t)=\sum_j A_j(\omega_j)\mathrm{e}^{i\omega_j t}$$

it is easy to calculate the output using the frequency response function as

$$y(t)=\sum_j A_j(\omega_j)H(\omega_j)\mathrm{e}^{i\omega_j t}$$

Thus a complicated convolution in the time domain, as in equation (9.12), reduces to a simple multiplication in the frequency domain, and we shall see that linear systems are often easier to study in the frequency domain.

   Returning to the definition of the frequency response function as given by equations (9.10) and (9.11), note that some authors define $H(\omega)$ for negative as well as positive frequencies. But for real-valued processes we need only consider $H(\omega)$ for $\omega>0$. Note that, in discrete time, $H(\omega)$ is only defined for frequencies up to the Nyquist frequency π (or $\pi/\Delta t$ if there is an interval Δt between successive observations). We have already introduced the Nyquist frequency in Section 7.2.1. Applying similar ideas to a linear system, it is clear that a sinusoidal input which has a higher frequency than π will have a corresponding sinusoid at a frequency in $(0, \pi)$ which gives identical readings at unit intervals of time and which will therefore give rise to an identical output.

   We have already noted that $H(\omega)$ is sometimes called the frequency response function and sometimes the transfer function. We will use the former term as it is more descriptive, indicating that the function shows how a linear system responds to sinusoids at different frequencies. In any case the term 'transfer function' is used by some authors in a different way. Engineers use the term to denote the Laplace transform of the impulse response function (see Appendix A). For a physically realizable stable system, the Fourier transform of the impulse response function may be regarded as a special case of the Laplace transform. A necessary and sufficient condition for a linear system to be stable is that the Laplace transform of the impulse response function should have no poles in the right half-plane or on the imaginary axis. For an unstable system, the Fourier transform does not exist, but the Laplace transform does. But we will only consider stable systems, in which case the Fourier transform is adequate. Note that Jenkins and Watts (1968) use the term 'transfer function' to denote the Z transform of the impulse response function (see Appendix A) in the discrete case. Z transforms are also used by engineers for discrete-time systems (e.g. Schwarz and Friedland, 1965).

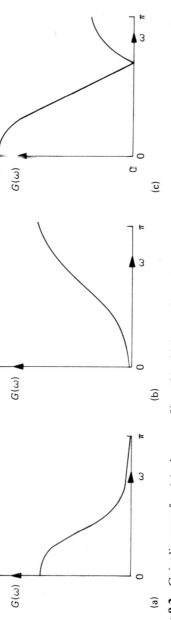

Figure 9.2 Gain diagrams for, (a) a low-pass filter; (b) a high-pass filter; (c) a simple moving average of three successive observations.

9.3.2   Gain and phase diagrams

The frequency response function $H(\omega)$ of a linear system is a complex function given by

$$H(\omega) = G(\omega)e^{i\phi(\omega)}$$

where $G(\omega)$, $\phi(\omega)$ are the gain and phase respectively. In order to understand the frequency properties of the system, it is useful to plot $G(\omega)$ and $\phi(\omega)$ against ω to obtain what are called the **gain diagram** and the **phase diagram**. If $G(\omega)$ is 'large' for low values of ω, but 'small' for high values of ω, as in Figure 9.2(a), then we have what is called a **low-pass** filter. This description is self-explanatory in that, if the input is a mixture of variation at several different frequencies, only those components with a low frequency will 'get through' the filter.

Conversely, if $G(\omega)$ is 'small' for low values of ω, but 'large' for high values of ω, then we have a **high-pass** filter as in Figure 9.2(b).

Plotting the phase diagram is complicated by the fact that the phase in equation (9.14) is not uniquely determined. If the gain is always taken to be positive, then the phase is undetermined by a multiple of 2π and is often constrained to the range $(-\pi, \pi)$. The (complex) value of $H(\omega)$ is examined to see which quadrant it is in. If $G(\omega)$ can be positive or negative, the phase is undetermined by a multiple of π and is often constrained to the range $(-\pi/2, \pi/2)$. These different conventions are not discussed adequately in many books (Hause, 1971, p. 214). In fact there are physical reasons why engineers prefer to plot the phase as a continuous unconstrained function, allowing $G(\omega)$ to be positive or negative, and using the fact that $\phi(0)=0$ provided $G(0)$ is finite.

9.3.3   Some examples

Example 9.1   Consider the simple moving average

$$y_t = (x_{t-1} + x_t + x_{t+1})/3$$

which is a linear system with impulse response function

$$h_k = \begin{cases} 1/3 & k = -1, 0, +1 \\ 0 & \text{otherwise} \end{cases}$$

The frequency response function of this filter is (using equation (9.11))

$$H(\omega) = \frac{1}{3}e^{-i\omega} + \frac{1}{3} + \frac{1}{3}e^{i\omega}$$

$$= \frac{1}{3} + \frac{2}{3}\cos\omega \qquad 0 < \omega < \pi$$

This function happens to be real, not complex, and so the phase appears to be given by

$$\phi(\omega)=0 \qquad 0<\omega<\pi$$

However, $H(\omega)$ is negative for $\omega>2\pi/3$, and so if we adopt the convention that the gain should be positive, then we have

$$G(\omega)=\left|\frac{1}{3}+\frac{2}{3}\cos\omega\right|$$

$$=\begin{cases} \dfrac{1}{3}+\dfrac{2}{3}\cos\omega & 0<\omega<2\pi/3 \\[2mm] -\dfrac{1}{3}-\dfrac{2}{3}\cos\omega & 2\pi/3<\omega<\pi \end{cases}$$

and

$$\phi(\omega)=\begin{cases} 0 & 0<\omega<2\pi/3 \\ \pi & 2\pi/3<\omega<\pi \end{cases}$$

The gain is plotted in Figure 9.2(c) and is of low-pass type. This is to be expected as a moving average smooths out local fluctuations (high-frequency variation) and measures the trend (the low-frequency variation). In fact it is probably more sensible to allow the gain to go negative in $(2\pi/3, \pi)$ so that the phase is zero for all ω in $(0, \pi)$.

Example 9.2 A linear system showing simple exponential response has impulse response function

$$h(u)=g\,e^{-u/T}/T \qquad u>0$$

Using (9.10), the frequency response function is

$$H(\omega)=g(1-i\omega T)/(1+\omega^2 T^2) \qquad \omega>0$$

Hence

$$G(\omega)=g/\sqrt{(1+\omega^2 T^2)}$$

$$\tan\phi(\omega)=-T\omega$$

As the frequency increases, $G(\omega)$ decreases so that the system is of low-pass type. As regards the phase, if we take $\phi(\omega)$ to be zero at zero frequency, then the phase becomes increasingly negative as ω increases until the output is out of phase with the input (see Figure 9.3).

Example 9.3 Consider the linear system consisting of pure delay, so that

$$y(t)=x(t-\tau)$$

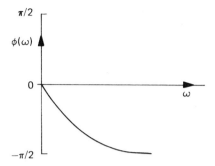

Figure 9.3 Phase diagram for a simple exponential response system.

where τ is a constant. Jenkins and Watts (1968) give the impulse response function as

$$h(u) = \delta(u - \tau)$$

where δ denotes the Dirac delta function (see Appendix B). Then the frequency response function is given by

$$H(\omega) = \int_{-\infty}^{\infty} \delta(u - \tau) e^{-i\omega u} \, du$$

$$= e^{-i\omega \tau}$$

In fact $H(\omega)$ can be derived without bringing in delta functions by using Theorem 9.1. Suppose that input $x(t) = e^{i\omega t}$ is applied to the system. Then the output is $y(t) = e^{i\omega(t-\tau)} = e^{-i\omega\tau} \times$ input. Thus, by analogy with equation (9.18), we have $H(\omega) = e^{-i\omega\tau}$.

For this linear system, the gain is constant, namely

$$G(\omega) = 1$$

while the phase is given by

$$\phi(\omega) = -\omega\tau$$

9.3.4 General relation between input and output

We have, so far, considered only sinusoidal inputs in the frequency domain. In this section we consider any type of input and show that it is generally easier to work with linear systems in the frequency domain than in the time domain.

The general relation between input and output in the time domain is given by equation (9.1), namely

$$y(t) = \int_{-\infty}^{\infty} h(u)x(t-u) \, du \qquad (9.19)$$

When $x(t)$ is not of a simple form, this integral may be hard to evaluate. Now consider the Fourier transform of the output, given by

$$Y(\omega) = \int_{-\infty}^{\infty} y(t)e^{-i\omega t}\, dt$$

$$= \int_{-\infty}^{\infty} \int_{-\infty}^{\infty} h(u)x(t-u)e^{-i\omega t}\, du\, dt$$

$$= \int_{-\infty}^{\infty} \int_{-\infty}^{\infty} h(u)e^{-i\omega u}x(t-u)e^{-i\omega(t-u)}\, du\, dt$$

But

$$\int_{-\infty}^{\infty} x(t-u)e^{-i\omega(t-u)}\, dt = \int_{-\infty}^{\infty} x(t)e^{-i\omega t}\, dt$$

for all values of u, and is therefore the Fourier transform of $x(t)$ which we will denote by $X(\omega)$. And

$$\int_{-\infty}^{\infty} h(u)e^{-i\omega u}\, du = H(\omega)$$

so that

$$Y(\omega) = H(\omega)X(\omega) \tag{9.20}$$

Thus the integral in (9.19) corresponds to a multiplication in the frequency domain provided that the Fourier transforms exist. A similar result holds in discrete time.

A more useful general relation between input and output, akin to equation (9.20), can be obtained when the input $x(t)$ is a stationary process with a continuous power spectrum. This result will be given as Theorem 9.2.

Theorem 9.2   Consider a stable linear system with gain function $G(\omega)$. Suppose that the input $X(t)$ is a stationary process with continuous power spectrum $f_x(\omega)$. Then the output $Y(t)$ is also a stationary process, whose power spectrum $f_y(\omega)$ is given by

$$f_y(\omega) = G^2(\omega)f_x(\omega) \tag{9.21}$$

Proof   The proof will be given for continuous time, but the same result holds for discrete time. Let us denote the impulse response and frequency response functions of the system by $h(u)$, $H(\omega)$ respectively. Thus $G(\omega) = |H(\omega)|$.

It is easy to show that a stationary input to a stable linear system gives rise to a stationary output, and this will not be shown here. For mathematical convenience, let us assume that the input has mean zero. Then the output

$$Y(t) = \int_{-\infty}^{\infty} h(u)X(t-u) \, du$$

also has mean zero.

Denote the autocovariance functions of $X(t)$, $Y(t)$ by $\gamma_x(\tau)$, $\gamma_y(\tau)$ respectively. Then

$$\gamma_y(\tau) = E[Y(t)Y(t+\tau)] \qquad \text{since } E[Y(t)] = 0$$

$$= E\left[\int_{-\infty}^{\infty} h(u)X(t-u) \, du \int_{-\infty}^{\infty} h(u')X(t+\tau-u') \, du'\right]$$

$$= \int\int_{-\infty}^{\infty} h(u)h(u')E[X(t-u)X(t+\tau-u')] \, du \, du'$$

But

$$E[X(t-u)X(t+\tau-u')] = \gamma_x(\tau-u'+u)$$

Thus

$$\gamma_y(\tau) = \int\int_{-\infty}^{\infty} h(u)h(u')\gamma_x(\tau-u'+u) \, du \, du' \qquad (9.22)$$

The relationship (9.22) between the autocovariance functions of input and output is not of a simple form. However, if we take Fourier transforms of both sides of (9.22) by multiplying by $e^{-i\omega\tau}/\pi$ and integrating with respect to τ, we find for the left-hand side

$$\frac{1}{\pi} \int_{-\infty}^{\infty} \gamma_y(\tau)e^{-i\omega t} \, d\tau = f_y(\omega)$$

from equation (6.17), while for the right-hand side

$$\int\int_{-\infty}^{\infty} h(u)h(u')\left[\frac{1}{\pi} \int_{-\infty}^{\infty} \gamma_x(\tau-u'+u)e^{-i\omega\tau}d\tau\right] du \, du'$$

$$= \int\int_{-\infty}^{\infty} h(u)e^{i\omega u}h(u')e^{-i\omega u'}\left[\frac{1}{\pi} \int_{-\infty}^{\infty} \gamma_x(\tau-u'+u)e^{-i\omega(\tau-u'+u)} \, d\tau\right] du \, du'$$

But

$$\frac{1}{\pi} \int_{-\infty}^{\infty} \gamma_x(\tau-u'+u)e^{-i\omega(\tau-u'+u)} \, d\tau = \frac{1}{\pi} \int_{-\infty}^{\infty} \gamma_x(\tau)e^{-i\omega\tau} \, d\tau = f_x(\omega)$$

for all u, u', and

$$\int_{-\infty}^{\infty} h(u)e^{i\omega u} = \overline{H(\omega)}$$

$$= G(\omega)e^{-i\phi(\omega)}$$

Thus

$$f_y(\omega) = \overline{H(\omega)}H(\omega)f_x(\omega)$$
$$= G^2(\omega)f_x(\omega)$$

This completes the proof.

The relationship between the spectra of the input and the output of a linear system is a very simple one. Once again a result in the frequency domain (equation (9.21)) is much simpler than the corresponding result in the time domain (equation (9.22)).

Theorem 9.2 can be used to evaluate the spectrum of some types of stationary process in a simpler way to that used in Chapter 6, where the method was to evaluate the autocovariance function of the process and then find its Fourier transform. Several examples will now be given.

(a) *Moving average process*

An MA process of order q is given by

$$X_t = \beta_0 Z_t + \cdots + \beta_q Z_{t-q}$$

where Z_t denotes a purely random process with variance σ_Z^2. This equation may be regarded as specifying a linear system with $\{Z_t\}$ as input and $\{X_t\}$ as output, whose frequency response function is given by

$$H(\omega) = \sum_{j=0}^{q} \beta_j e^{-i\omega j}$$

As $\{Z_t\}$ is a purely random process, its spectrum is given by (see equation (6.19))

$$f_Z(\omega) = \sigma_Z^2/\pi$$

Thus, using (9.21), the spectrum of $\{X_t\}$ is given by

$$f_x(\omega) = \left| \sum_{j=0}^{q} \beta_j e^{-i\omega j} \right|^2 \sigma_Z^2/\pi$$

For example, for the first-order MA process

$$X_t = Z_t + \beta Z_{t-1} \tag{9.23}$$

we have

$$H(\omega) = 1 + \beta e^{-i\omega}$$

and

$$G^2(\omega) = |H(\omega)|^2 = (1 + \beta \cos \omega)^2 + \beta^2 \sin^2 \omega$$
$$= 1 + 2\beta \cos \omega + \beta^2$$

so that $f_x(\omega) = (1 + 2\beta \cos \omega + \beta^2)\sigma_Z^2/\pi$ as already derived in Section 6.5.

This type of approach can also be used when $\{Z_t\}$ is not a purely random process. For example, suppose that the $\{Z_t\}$ process in equation (9.23) has arbitrary spectrum $f_Z(\omega)$. Then the spectrum of $\{X_t\}$ is given by

$$f_x(\omega) = (1 + 2\beta \cos \omega + \beta^2)f_Z(\omega)$$

(b) Autoregressive process

The first-order AR process

$$X_t = \alpha X_{t-1} + Z_t$$

may be regarded as a linear system producing output X_t from input Z_t. It may also be regarded as a linear system 'the other way round', producing output Z_t from input X_t by

$$Z_t = X_t - \alpha X_{t-1}$$

This formulation has frequency response function

$$H(\omega) = 1 - \alpha e^{-i\omega}$$

and gives the desired result in a mathematically simpler way. Thus

$$G^2(\omega) = 1 - 2\alpha \cos \omega + \alpha^2$$

and so

$$f_Z(\omega) = (1 - 2\alpha \cos \omega + \alpha^2)f_x(\omega) \qquad (9.24)$$

But if $\{Z_t\}$ denotes a purely random process with spectrum $f_Z(\omega) = \sigma_Z^2/\pi$, then equation (9.24) may be rewritten to evaluate $f_x(\omega)$ as

$$f_x(\omega) = \sigma_Z^2/\pi(1 - 2\alpha \cos \omega + \alpha^2)$$

which has already been obtained as equation (6.23) by the earlier method. This approach may also be used for higher-order AR processes.

(c) Differentiation

Consider the linear system which converts a continuous input $X(t)$ into output $Y(t)$ by

$$Y(t) = \frac{dX(t)}{dt} \qquad (9.25)$$

A differentiator is of considerable mathematical interest, although in practice only approximations to it are physically realizable.

If the input is sinusoidal, $X(t) = e^{i\omega t}$, then the output is given by

$$Y(t) = i\omega\, e^{i\omega t}$$

so that, using (9.18), the frequency response function is

$$H(\omega) = i\omega$$

If the input is a stationary process, with spectrum $f_x(\omega)$, then it appears that the output has spectrum

$$f_y(\omega) = |i\omega|^2 f_x(\omega)$$
$$= \omega^2 f_x(\omega) \tag{9.26}$$

However, this result assumes that the linear system (9.25) is stable, when in fact it is only stable for certain types of input process. For example, it is clear that the response to a unit step change is an unbounded impulse. In order for the system to be stable, the variance of the output must be finite. Now

$$\text{var}[Y(t)] = \int_0^\infty f_y(\omega)\, d\omega$$
$$= \int_0^\infty \omega^2 f_x(\omega)\, d\omega$$

But, using equation (6.18), we have

$$\gamma_x(k) = \int_0^\infty f_x(\omega)\cos \omega k\, d\omega$$

and

$$\frac{d^2\gamma_x(k)}{dk^2} = -\int_0^\infty \omega^2 f_x(\omega)\cos \omega k\, d\omega$$

so that

$$\text{Var}[Y(t)] = -\left[\frac{d^2\gamma_x(k)}{dk^2}\right]_{k=0}$$

Thus $Y(t)$ has finite variance provided that $\gamma_x(k)$ can be differentiated twice at $k=0$, and only then does equation (9.26) hold.

9.3.5   Linear systems in series

The advantages of working in the frequency domain are also evident when we consider two or more linear systems in series (or in cascade). For example Figure 9.4 shows two linear systems in series, where the input $x(t)$ to system I produces output $y(t)$ which in turn is the input to system II producing output $z(t)$. It is often of interest to evaluate the properties of the overall system, which is also linear, where $x(t)$ is the input and $z(t)$ is the output. We will denote the

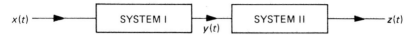

Figure 9.4   Two linear systems in series.

impulse response and frequency response functions of systems I and II by $h_1(u)$, $h_2(u)$, $H_1(\omega)$ and $H_2(\omega)$.

In the time domain, the relationship between $x(t)$ and $z(t)$ would be in the form of a double integral involving $h_1(u)$ and $h_2(u)$, which is rather complicated. But in the frequency domain we can denote the Fourier transforms of $x(t)$, $y(t)$, $z(t)$ by $X(\omega)$, $Y(\omega)$, $Z(\omega)$ and use equation (9.20). Then

$$Y(\omega) = H_1(\omega)X(\omega)$$

and

$$Z(\omega) = H_2(\omega)Y(\omega)$$
$$= H_2(\omega)H_1(\omega)X(\omega)$$

Thus it is easy to see that the overall frequency response function of the combined system is

$$H(\omega) = H_1(\omega)H_2(\omega) \tag{9.27}$$

If

$$H_1(\omega) = G_1(\omega)e^{i\phi_1(\omega)}$$
$$H_2(\omega) = G_2(\omega)e^{i\phi_2(\omega)}$$

then

$$H(\omega) = G_1(\omega)G_2(\omega)e^{i[\phi_1(\omega) + \phi_2(\omega)]}$$

Thus the overall gain is the **product** of the component gains, while the overall phase is the **sum** of the component phases.

The above results are easily extended to the situation where there are k linear systems in series with respective frequency response functions $H_1(\omega), \ldots, H_k(\omega)$. The overall frequency response function is

$$H(\omega) = H_1(\omega)H_2(\omega) \ldots H_k(\omega)$$

9.3.6   Design of filters

The results of this section enable us to consider in more depth the properties of the filters introduced in Section 2.5.2. Given a time series $\{x_t\}$, the filters for estimating or removing trend are of the form

$$y_t = \sum_k h_k x_{t-k}$$

and are clearly linear systems with frequency response function

$$H(\omega)=\sum_{k} h_{k}\, e^{-i\omega k}$$

If the time series has spectrum $f_{x}(\omega)$, then the spectrum of the smoothed series is given by

$$f_{y}(\omega)=G^{2}(\omega)f_{x}(\omega) \tag{9.28}$$

where $G(\omega)=|H(\omega)|$.

How do we set about choosing an appropriate filter for a time series? The design of a filter involves a choice of $\{h_{k}\}$ and hence of $H(\omega)$ and $G(\omega)$. Two types of 'ideal' filter are shown in Figure 9.5. Both have sharp cut-offs, the low-pass filter completely eliminating high-frequency variation and the high-pass filter completely eliminating low-frequency variation.

But 'ideal' filters of this type are impossible to achieve with a finite set of weights. Instead the smaller the number of weights used, the less sharp will generally be the cut-off property of the filter. For example the gain diagram of a simple moving average of three successive observations (Figure 9.2(c)) is of low-pass type but has a much less sharp cut-off than the 'ideal' low-pass filter (Figure 9.5(a)). More sophisticated moving averages such as Spencer's 15-point moving average have much better cut-off properties.

The differencing filter (Section 2.5.3), for removing a trend, of the form

$$y_{t}=x_{t}-x_{t-1}$$

has frequency response function

$$H(\omega)=1-e^{-i\omega}$$

and gain function

$$G(\omega)=\sqrt{[2(1-\cos\omega)]}$$

which is plotted in Figure 9.6. This is indeed of high-pass type, but the cut-off

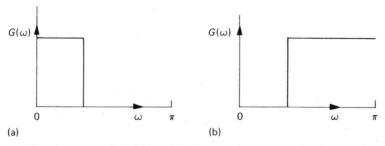

(a)   (b)

Figure 9.5   Two types of ideal filter, (a) a low-pass filter or trend estimator; (b) a high-pass filter or trend eliminator.

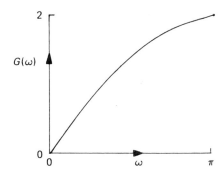

Figure 9.6   The gain diagram for the difference operator.

property is rather poor and this should be borne in mind when working with first differences.

9.4   IDENTIFICATION OF LINEAR SYSTEMS

We have so far assumed that the structure of the linear system under consideration is known. Given the impulse response function of a system, or equivalently the frequency response function, we can find the output corresponding to a given input. In particular, when considering the properties of filters for estimating trend and seasonality, a formula for the 'system' is given.

But many problems concerning linear systems are of a completely different type. The structure of the system is **not** known and the problem is to examine the relationship between input and output so as to infer the properties of the system. This procedure is called the **identification** of the system. For example, suppose we are interested in the effect of temperature on the yield from a chemical process. Here we have a physical system which we assume, initially at least, is approximately linear over the range of interest. By examining the relationship between observations on temperature (the input) and yield (the output) we can infer the properties of the chemical process.

The identification process is straightforward if the input to the system can be controlled and if the system is 'not contaminated by noise'. In this case, we can simply apply an impulse or step change input, observe the output, and hence estimate the impulse response or step response function. Alternatively we can apply sinusoidal inputs at different frequencies and observe the amplitude and phase shift of the corresponding sinusoidal outputs. This enables us to evaluate the gain and phase diagrams.

But many systems are contaminated by noise as illustrated in Figure 9.7, where $N(t)$ denotes a noise process. This noise process may not be white noise

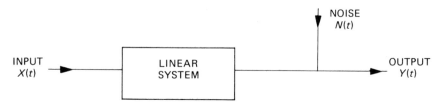

Figure 9.7   A linear system with added noise.

(i.e. may not be a purely random process), but is usually assumed to be uncorrelated with the input process $X(t)$.

A further difficulty arises when the input is observable but not controllable. In other words one cannot make changes, such as a step change, to the input. For example, attempts have been made to treat the economy as a linear system and to examine the relationship between variables like price increases (input) and wage increases (output). But price increases can only be controlled to a certain extent by governmental decisions, and there is also a feedback problem in that wage increases may in turn affect price increases (see Section 9.4.3).

When the system is affected by noise and/or the input is not controllable, more refined techniques are required to identify the system. We will describe two alternative approaches, one in the frequency domain and one in the time domain. In Section 9.4.1 we show how cross-spectral analysis of input and output may be used to estimate the frequency response function of a linear system. In Section 9.4.2 we describe a method proposed by Box and Jenkins (1970) for estimating the impulse response function of a linear system.

9.4.1   Estimating the frequency response function

Suppose that we have a linear system with added noise, as depicted in Figure 9.7, where the noise is assumed to be uncorrelated with the input and to have mean zero. Suppose also that we have observations on input and output over some time period and wish to estimate the frequency response function of the system. We will denote the (unknown) impulse response and frequency response functions of the system by $h(u)$, $H(\omega)$ respectively.

The reader may think that equation (9.21), namely

$$f_y(\omega) = G^2(\omega) f_x(\omega)$$

can be used to estimate the gain of the system. But this equation does not hold in the presence of noise $N(t)$, and does not in any case give information about the phase of the system. Instead we will derive a relationship involving the cross-spectrum of input and output.

In continuous time, the output $Y(t)$ is given by

$$Y(t) = \int_0^\infty h(u)X(t-u)\,du + N(t) \tag{9.29}$$

Note that we are only considering physically realizable systems, so that $h(u)$ is zero for $u<0$. For mathematical convenience we assume $E[X(t)]=0$ so that $E[Y(t)]=0$, but the following results also hold if $E[X(t)]\neq0$. Multiplying through equation (9.29) by $X(t-\tau)$ and taking expectations, we have

$$E[N(t)X(t-\tau)]=0$$

since $N(t)$ is assumed to be uncorrelated with input, so that

$$\gamma_{xy}(\tau) = \int_0^\infty h(u)\gamma_{xx}(\tau-u)\,du \tag{9.30}$$

where γ_{xy} is the cross-covariance function of $X(t)$ and $Y(t)$, and γ_{xx} is the autocovariance function of $X(t)$. Equation (9.30) is called the Wiener-Hopf integral equation and, given γ_{xy} and γ_{xx}, can in principle be solved to find the impulse response function $h(u)$. But it is often easier to work with the corresponding relationship in the frequency domain.

First we revert to discrete time and note that the discrete-time analogue of equation (9.30) is

$$\gamma_{xy}(\tau) = \sum_{k=0}^\infty h_k \gamma_{xx}(\tau-k) \tag{9.31}$$

Take Fourier transforms of both sides of this equation by multiplying by $e^{-i\omega\tau}/\pi$ and summing from $\tau = -\infty$ to $+\infty$. Then we find

$$f_{xy}(\omega) = \sum_{\tau=-\infty}^\infty \sum_{k=0}^\infty h_k\, e^{-i\omega k}\, \gamma_{xx}(\tau-k)e^{-i\omega(\tau-k)}/\pi$$

$$= \sum_{k=0}^\infty h_k\, e^{-i\omega k}\, f_x(\omega)$$

$$= H(\omega)f_x(\omega) \tag{9.32}$$

where f_{xy} is the cross-spectrum of input and output and f_x is the (auto)spectrum of the input. Thus, once again, a convolution in the time domain corresponds to a multiplication in the frequency domain.

Estimates of $f_{xy}(\omega)$ and $f_x(\omega)$ can now be used to estimate $H(\omega)$ using (9.32). Denote the estimated spectrum of the input by $\hat{f}_x(\omega)$, and the estimate of f_{xy}, obtained by cross-spectral analysis, by $\hat{f}_{xy}(\omega)$. Then

$$\hat{H}(\omega) = \hat{f}_{xy}(\omega)/\hat{f}_x(\omega)$$

We usually write

$$H(\omega) = G(\omega)e^{i\phi(\omega)}$$

and estimate the gain and phase separately. We have

$$\hat{G}(\omega) = |\hat{H}(\omega)| = |\hat{f}_{xy}(\omega)/\hat{f}_x(\omega)|$$

$$= |\hat{f}_{xy}(\omega)|/\hat{f}_x(\omega) \qquad \text{since } f_x(\omega) \text{ is real}$$

$$= \hat{\alpha}_{xy}(\omega)/\hat{f}_x(\omega) \tag{9.33}$$

where $\alpha_{xy}(\omega)$ is the cross-amplitude spectrum (see equation (8.13)).
We also find

$$\tan \hat{\phi}(\omega) = -\hat{q}(\omega)/\hat{c}(\omega) \tag{9.34}$$

where $q(\omega)$, $c(\omega)$ are the quadrature and co-spectra, respectively (see equation (8.14)).

Thus, having estimated the cross-spectrum, equations (9.33) and (9.34) enable us to estimate the gain and phase of the linear system, whether or not there is added noise.

We can also use cross-spectral analysis to estimate the properties of the noise process. The discrete-time version of equation (9.29) is

$$Y_t = \sum_{k=0}^{\infty} h_k X_{t-k} + N_t \tag{9.35}$$

For mathematical convenience, we again assume that $E(N_t) = E(X_t) = 0$ so that $E(Y_t) = 0$. If we multiply both sides of (9.35) by Y_{t-m}, we find

$$Y_t Y_{t-m} = (\Sigma h_k X_{t-k} + N_t)(\Sigma h_k X_{t-m-k} + N_{t-m})$$

Taking expectations we find

$$\gamma_{yy}(m) = \sum_k \sum_j h_k h_j \gamma_{xx}(m-k+j) + \gamma_{nn}(m)$$

since $\{X_t\}$ and $\{N_t\}$ are assumed to be uncorrelated. Taking Fourier transforms of both sides of this equation, we find

$$f_y(\omega) = H(\omega)\overline{H(\omega)}f_x(\omega) + f_n(\omega)$$

But

$$H(\omega)\overline{H(\omega)} = G^2(\omega)$$

$$= C(\omega)f_y(\omega)/f_x(\omega)$$

so that

$$f_n(\omega) = f_y(\omega)[1 - C(\omega)] \tag{9.36}$$

Thus an estimate of $f_n(\omega)$ is given by

$$\hat{f}_n(\omega) = \hat{f}_y(\omega)[1 - \hat{C}(\omega)]$$

Equation (9.36) also enables us to see that if there is no noise, so that there is a pure linear relation between X_t and Y_t, then $f_n(\omega)=0$ and $C(\omega)=1$ for all ω. On the other hand if $C(\omega)=0$ for all ω, then $f_y(\omega)=f_n(\omega)$ and the output is not linearly related to the input. This confirms the point mentioned in Chapter 8 that the coherency $C(\omega)$ measures the linear correlation between input and output at frequency ω.

The results of this section not only show us how to identify a linear system by cross-spectral analysis, but also give further guidance on the interpretation of functions derived from the cross-spectrum, particularly the gain, phase and coherency. An example, involving a chemical process, is given by Goodman *et al.* (1961), while Gudmundsson (1971) describes an economic applicaiton of cross-spectral analysis.

In principle, estimates of the frequency response function of a linear system may be transformed to give estimates of the impulse response function (Jenkins and Watts, 1968, p. 444) but I do not recommend this. For instance Example 8.4 appears to indicate that the sign of the phase may be used to indicate which series is 'leading' the other. But Hause (1971) has shown that for more complicated lagged models of the form

$$Y_t = \sum_{k=0}^{\infty} h_k X_{t-k} \tag{9.37}$$

which are called **distributed lag** models by economists, it becomes increasingly difficult to make inferences from phase estimates. Hause concludes that phase leads and lags will rarely provide economists with direct estimates of the time-domain relationships that are of more interest to them.

9.4.2 The Box-Jenkins approach

This section gives a brief introduction to the method proposed by G. E. P. Box and G. M. Jenkins for identifying a physically realizable linear system, in the time domain, in the presence of added noise. Further details can be found in Box and Jenkins (1968, 1970).

The input and output series are both differenced d times until both are stationary, and are also mean-corrected. The modified series will be denoted by $\{X_t\}, \{Y_t\}$, respectively. We want to find the impulse response function $\{h_k\}$ of the system, where

$$Y_t = \sum_{k=0}^{\infty} h_k X_{t-k} + N_t \tag{9.38}$$

The 'obvious' way to estimate $\{h_k\}$ is to multiply through equation (9.38) by X_{t-m} and take expectations to give

$$\gamma_{xy}(m) = h_0 \gamma_{xx}(m) + h_1 \gamma_{xx}(m-1) + \cdots \tag{9.39}$$

assuming that N_t is uncorrelated with the input. If we assume that the weights $\{h_k\}$ are effectively zero beyond $k = K$, then the first $K+1$ equations of type (9.39) for $m = 0, 1, \ldots, K$, can be solved for the $K+1$ unknowns $h_0, h_1, \ldots, h_K$, on substituting estimates of γ_{xy} and γ_{xx}. Unfortunately these equations do not, in general, provide good estimators for the $\{h_k\}$, and, in any case, assume knowledge of the truncation point K. The basic trouble, as already noted in Section 8.1.2, is that autocorrelation within the input and output series will increase the variance of cross-correlation estimates.

Box and Jenkins (1970) therefore propose two modifications to the above procedure. First, they suggest 'prewhitening' the input before calculating the sample cross-covariance function. Secondly, they propose an alternative form of equation (9.37) which will in general require fewer parameters. They represent the linear system by the equation

$$Y_t - \delta_1 Y_{t-1} - \cdots - \delta_r Y_{t-r}$$
$$= \omega_0 X_{t-b} - \omega_1 X_{t-b-1} - \cdots - \omega_s X_{t-b-s} \qquad (9.40)$$

This is rather like equation (9.4), but is given in the notation used by Box and Jenkins (1970) and involves an extra parameter b, which is called the **delay** of the system. The delay can be any non-negative integer. Using the backward shift operator B, (9.40) may be written as

$$\delta(B) Y_t = \omega(B) X_{t-b} \qquad (9.41)$$

where

$$\delta(B) = 1 - \delta_1 B - \cdots - \delta_r B^r$$

and

$$\omega(B) = \omega_0 - \omega_1 B - \cdots - \omega_s B^s$$

Box and Jenkins (1970) describe equation (9.41) as a **transfer function** model, which is a potentially misleading description in that the term 'transfer function' is often used to describe some sort of transform of the impulse response function.

The Box-Jenkins procedure begins by fitting an ARMA model to the (differenced) input. Suppose this model is of the form (see Section 3.4.5)

$$\phi(B) X_t = \theta(B) \alpha_t$$

where $\{\alpha_t\}$ denotes a purely random process, in the notation of Box and Jenkins. Thus we can transform the input to a white noise process by

$$\phi(B) \theta^{-1}(B) X_t = \alpha_t$$

Suppose we apply the same transformation to the output, to give

$$\phi(B) \theta^{-1}(B) Y_t = \beta_t$$

and then calculate the cross-covariance function of the filtered input and output, namely $\{\alpha_t\}$ and $\{\beta_t\}$. It turns out that this function gives a better estimate of the impulse response function, since if we write

$$h(B) = h_0 + h_1 B + h_2 B^2 + \cdots$$

so that

$$Y_t = h(B)X_t + N_t$$

then

$$\begin{aligned}
\beta_t &= \phi(B)\theta^{-1}(B)Y_t \\
&= \phi(B)\theta^{-1}(B)[h(B)X_t + N_t] \\
&= h(B)\alpha_t + \phi(B)\theta^{-1}(B)N_t
\end{aligned}$$

and

$$\gamma_{\alpha\beta}(m) = h_m \, \mathrm{Var}(\alpha_t) \tag{9.42}$$

since $\{\alpha_t\}$ is a purely random process, and N_t is uncorrelated with $\{\alpha_t\}$. Equation (9.42) is of a much simpler form to equation (9.39). If we denote the sample cross-variance function of α_t and β_t by $c_{\alpha\beta}$, and the observed variance of α_t by s_α^2, then an estimate of h_m is given by

$$\hat{h}_m = c_{\alpha\beta}(m)/s_\alpha^2 \tag{9.43}$$

These estimates should be more reliable than those given by the solution of equations of type (9.39).

Box and Jenkins (1970) give the theoretical impulse response functions for a variety of models of type (9.40) and go on to show how the shape of the estimated impulse response function given by (9.43) can be used to suggest appropriate values of the integers r, b and s in equation (9.40). They then show how to obtain least squares estimates of $\delta_1, \delta_2, \ldots, \omega_0, \omega_1, \ldots$, given values of r, b and s. These estimates can in turn be used to obtain refined estimates of $\{h_m\}$ if desired.

Box and Jenkins go on to show how a model of type (9.40), with added noise, can be used for forecasting and control. Some successful case studies have now been published (e.g. Jenkins, 1979; Jenkins and Mcleod, 1982) and the method looks potentially useful. However, it should be noted that the main example discussed by Box and Jenkins (1970), using some gas furnace data, has been criticized by Young (1974) and Chatfield (1977, p. 504) on a number of grounds, including the exceptionally high correlation between input and output which means that virtually any identification procedure will give good results.

A modification to the Box-Jenkins procedure has been proposed by Haugh and Box (1977) in which separate models are fitted to both the input and the output before cross-correlating the resulting residual series. Further experience is needed to see if this is generally advisable. The modified procedure has been used by Pierce (1977) to assess the relationship between various pairs of economic series, and he shows that, after taking account of autocorrelation within each series, there is usually very little correlation left between the residual series. Pierce discusses how to reconcile these results with the views of economists.

The question as to when it is better to use cross-spectral analysis or a time-domain parametric model approach is still unresolved, though it has been discussed by a number of authors (e.g. Astrom and Eykhoff, 1971). It seems to me to be unwise to attempt to make general pronouncements on the relative virtues of time-domain and frequency-domain methods. The two approaches appear to be complementary rather than rivals, and it may be helpful to try both.

A transfer function model may be regarded as a special case of a multivariate ARIMA model (see Section 11.9). If the open-loop nature of the data is open to question, perhaps because of the presence of feedback (see Section 9.4.3), then it may be advisable to try to fit the more general multivariate ARIMA model.

In conclusion, it is worth noting that a similar method to the Box-Jenkins approach has been independently developed by two control engineers, K. J. Astrom and T. Bohlin. This method also involves prewhitening and a model of type (9.40) (Astrom and Bohlin, 1966; Astrom, 1970), but does not discuss identification and estimation procedures in equal depth. One difference in the Astrom-Bohlin approach is that non-stationary series may be converted to stationarity by high-pass filtering methods other than differencing.

9.4.3   Systems involving feedback

A system of the type illustrated in Figure 9.7 is called an **open-loop** system, and the procedures described in the previous two sections are appropriate for data collected under these conditions. But data are often collected from systems where some form of **feedback control** is being applied, and then we have what is called a **closed-loop** system as illustrated in Figure 9.8. For example, when trying to identify a full-scale industrial process, it could be dangerous, or an unsatisfactory product could be produced, if some form of feedback control is not applied to keep the output somewhere near target. Similar problems arise in an economic context. For example, attempts to find a linear relationship showing the effect of price changes on wage changes are bedevilled by the fact that wage changes will in turn affect prices.

The problem of identifying systems in the presence of feedback control is discussed by Granger and Hatanaka (1964, Chapter 7), Astrom and Eykhoff

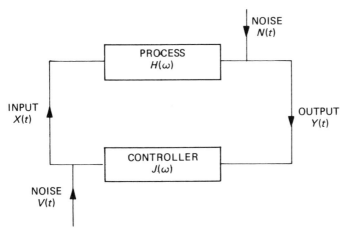

Figure 9.8 A closed-loop system.

(1971, p. 130), Gustavsson, Ljung and Soderstrom (1977) and Priestley (1983), and it is important to realize that open-loop procedures may not be applicable. The situation can be explained more clearly in the frequency domain. Let $f_{xy}(\omega)$ denote the cross-spectrum of $X(t)$ and $Y(t)$ in Figure 9.8, and let $f_x(\omega), f_n(\omega), f_v(\omega)$ denote the spectra of $X(t)$, $N(t)$ and $V(t)$ respectively. Then if $H(\omega)$ and $J(\omega)$ denote the frequency response functions of the system and controller respectively, it can be shown (e.g. Akaike, 1967) that

$$f_{xy}/f_x \quad (Hf_v \mid \bar{J}f_n)/(f_v + J\bar{J}f_n) \tag{9.44}$$

where all terms are functions of frequency, and $\bar{J}$ is the complex conjugate of J. Only if $f_n \equiv 0$ or $J \equiv 0$ is the ratio f_{xy}/f_x equal to H as is the case for an open-loop system (equation (9.32)). Thus the estimate of H provided by $\hat{f}_{xy}/\hat{f}_x$ will be poor unless f_n/f_v is small. In particular, if $f_v \equiv 0$, $\hat{f}_{xy}/\hat{f}_x$ will provide an estimate of J^{-1} and **not** of H.

Similar remarks apply to an analysis in the time domain. The time-domain equivalent of (9.44) is given by Box and MacGregor (1974).

The above problem is not specifically discussed by Box and Jenkins (1970), although it is quite clear from the remarks in their Section 11.6 that their methods are only intended for use in open-loop systems. However, some confusion may be created by the fact that Box and Jenkins (1970, Section 12.2) do discuss ways of choosing optimal feedback control, which is quite a different problem. Having identified a system in open loop, they show how to choose the feedback control action so as to satisfy some chosen criterion.

Unfortunately, open-loop identification procedures have sometimes been used for a closed-loop system where they are not appropriate. Tee and Wu (1972) studied a paper machine while it was already operating under manual control and proposed a control procedure which has been shown to be worse than the existing form of control (Box and MacGregor, 1974). In marketing, several authors have investigated the relationship between expenditure on offers and advertising on the sales of products such as washing-up liquid and coffee. However, expenditure on advertising is often in turn affected by changes in sales levels, so that any conclusions obtained by an open-loop analysis are open to doubt.

What then can be done if feedback is present? Box and MacGregor (1974) suggest one possible approach in which one deliberately adds an independent programmed noise sequence on top of the noise $V(t)$. Alternatively one may have some knowledge of the noise structure or of the controller frequency response function. Akaike (1968) claims that it is possible to identify a system provided only that instantaneous transmission of information does not occur in both system and controller, and an example of his, rather complicated, procedure is given by Otomo, Nakagawa and Akaike (1972).

However, a further difficulty is that it is not always clear if feedback is present or not, particularly in economics and marketing. Some indication may be given by the methods of Granger and Hatanaka (1964, Chapter 7), or by observing significantly large cross-correlation coefficients between (pre-whitened) input and output at a zero or positive lag. Alternatively it may be clear from physical considerations that feedback is or is not present. For example in studying the relationship between average (ambient) temperature and sales of a product, it is clear that sales cannot possibly affect temperature so that one has an open-loop system. However, if feedback is present, then it is important to realize that cross-correlation and cross-spectral analysis of the raw data may give misleading results.

EXERCISES

9.1 Find the impulse response function, the step response function, the frequency response function, the gain and the phase shift for the following linear systems (or filters):

(a) $y_t = \frac{1}{2}x_{t-1} + x_t + \frac{1}{2}x_{t+1}$

(b) $y_t = \frac{1}{5}(x_{t-2} + x_{t-1} + x_t + x_{t+1} + x_{t+2})$

(c) $y_t = \nabla x_t$

(d) $y_t = \nabla^2 x_t$

where in each case t is integer-valued. Plot the gain and phase shift for filters (a) and (c). Which of the filters are low-pass and which high-pass?

If filters (a) and (b) are joined in series, find the frequency response function of the combined filter.

9.2 Find the frequency response functions of the following linear systems in continuous time:

(a) $y(t) = gx(t - \tau)$

(b) $y(t) = \dfrac{g}{T} \displaystyle\int_0^\infty e^{-u/T} x(t - u)\, du$

where g, T and τ are positive constants.

9.3 If $\{X_t\}$ is a stationary discrete time series with power spectral density function $f(\omega)$, show that the smoothed time series

$$Y_t = \sum_{p=0}^{k} a_p X_{t-p}$$

where the as are real constants, is a stationary process which has power spectral density function

$$\left[\sum_{q=0}^{k} \sum_{p=0}^{k} a_p a_q \cos(p - q)\omega \right] f(\omega)$$

In particular, if $a_p = 1/(k+1)$ for $p = 0, 1, \ldots, k$, show that the power spectrum of Y_t is

$$f(\omega)[1 - \cos(k+1)\omega]/(k+1)^2(1 - \cos\omega)$$

(Hint: Use equation (9.21) and the trigonometric relation $\cos A \cos B = \frac{1}{2}[\cos(A+B) + \cos(A-B)]$.)

9.4 Consider the one-parameter second-order AR process

$$X_t = \alpha X_{t-2} + Z_t$$

where $\{Z_t\}$ denotes a purely random process, mean zero, variance σ_Z^2. Show that the process is second-order stationary if $|\alpha| < 1$, and find its autocovariance and autocorrelation functions.

Show that the power spectral density function of the process is given by

$$f(\omega) = \sigma_Z^2/\pi(1 - 2\alpha \cos 2\omega + \alpha^2) \qquad 0 < \omega < \pi$$

using two different methods: (a) by transforming the autocovariance function; (b) by using the approach of Section 9.3.4.

Suppose now that $\{Z_t\}$ is any stationary process which has power spectrum $f_Z(\omega)$. What then is the power spectrum of $\{X_t\}$ as defined by the above equation?

9.5 Show that the power spectral density function of the ARMA(1, 1) process

$$X_t = \alpha X_{t-1} + Z_t + \beta Z_{t-1}$$

is given by $f_X(\omega)=\sigma_Z^2(1+2\beta\cos\omega+\beta^2)/\pi(1-2\alpha\cos\omega+\alpha^2)$ for $0<\omega<\pi$, using the approach of Section 9.3.4. It may help to let $Y_t=Z_t+\beta Z_{t-1}$. (This power spectrum may be shown to be equivalent to the normalized spectrum in Exercise 6.7 after some algebra.)

More generally, for the MA process $X_t=\theta(B)Z_t$, show that the transfer function of the filter $Z_t\to X_t$ is $H(\omega)=\theta(e^{-i\omega})$, so that the spectrum of X_t is given by $\theta(e^{-i\omega})\theta(e^{i\omega})\sigma_Z^2/\pi$ for $0<\omega<\pi$. Hence show that the spectrum of the general ARMA process $\phi(B)X_t=\theta(B)Z_t$ is given by $\theta(e^{-i\omega})\theta(e^{i\omega})\sigma_Z^2/\pi\phi(e^{-i\omega})\phi(e^{i\omega})$.

10
State-space models and the Kalman filter

A general class of models, arousing much current interest, is that of state-space models. They were originally developed by control engineers, particularly for applications concerning navigation systems such as controlling the position of a space rocket. However, they have also been found to be useful in many types of time-series problem, such as short-term forecasting.

This chapter introduces state-space models for the time-series analyst, as well as describing the Kalman filter, which is an important general method of handling state-space models. Essentially, Kalman filtering is a method of signal processing which gives optimal estimates of the current state of a dynamic system. It consists of a set of equations for recursively estimating the current state of a system and for finding variances of these estimates. An alternative clear introduction is given by Harvey (1984), while an engineering viewpoint is given for example by Anderson and Moore (1979) and Mayhook (1979).

10.1 STATE-SPACE MODELS

When the scientist tries to measure any sort of signal it will typically be contaminated by noise, so that the actual observation X_t is given (in words) by

$$observation = signal + noise \tag{10.1}$$

In state-space models the signal is taken to be a linear combination of a set of variables, called **state variables**, which constitute what is called the **state vector** at time t. This vector describes the state of the system at time t, and is sometimes called the 'state of nature'.

We have used the jargon of control engineering, but the ideas are equally applicable to statistical problems. For example the observation could be some observed economic variable, and the state variables could then include such quantities as the current true underlying level and the current seasonal factor (if any).

It is an unfortunate complication that there is no standard notation for the problem. We denote the $(m \times 1)$ state vector by $\boldsymbol{\theta}_t$, and write equation (10.1) as

$$X_t = \mathbf{h}_t^{\mathrm{T}} \boldsymbol{\theta}_t + n_t \qquad (10.2)$$

where the $(m \times 1)$ vector $\mathbf{h}_t$ is assumed to be a known vector and n_t denotes the observation error.

The state vector $\boldsymbol{\theta}_t$, which is of prime importance, cannot be observed directly (i.e. is unobservable), and so we wish to use the observations on X_t to make inferences about $\boldsymbol{\theta}_t$, Although not directly observable, it is often reasonable to assume that we know how $\boldsymbol{\theta}_t$ changes through time, and we denote the updating equation by

$$\boldsymbol{\theta}_t = G_t \boldsymbol{\theta}_{t-1} + \mathbf{w}_t \qquad (10.3)$$

where the $(m \times m)$ matrix G_t is assumed known, and $\mathbf{w}_t$ denotes a vector of deviations.

The two equations (10.2) and (10.3) constitute the general form of the (univariate) state-space model. Equation (10.2) is called the **observation** (or **measurement**) equation, while (10.3) is called the **transition** (or **system**) equation.

The 'errors' in the observation and transition equations are generally assumed to be uncorrelated with each other at all time periods, and also to be serially uncorrelated. We may further assume that n_t is $N(0, \sigma_n^2)$ while $\mathbf{w}_t$ is multivariate normal with zero mean vector and known variance-covariance matrix denoted by W_t.

The state-space model can readily be generalized to the case where X_t is a vector by making $\mathbf{h}_t$ a matrix of appropriate size and by making n_t a vector of appropriate length. It is also possible to add terms involving known linear combinations of explanatory (or exogenous) variables to the right-hand side of (10.2).

The application of state-space models to engineering problems, such as controlling a dynamic system, is fairly clear. There the equations of motion of a system are often assumed to be known *a priori*, as are the properties of the system disturbances and measurement errors, although some model parameters may have to be estimated from data. Neither the equations nor the 'error' statistics need be constant as long as they are known functions of time. However, at first sight, state-space models may appear to have little connection with earlier time-series models. Nevertheless it can be shown, for example, that it is possible to put many types of time-series model into the state-space formulation. They include regression and ARMA models as well as the sort of trend-and-seasonal model for which exponential smoothing methods are thought to be appropriate. Bayesian forecasting (see Section 10.1.5) also relies on what is essentially a state-space representation, while some models with time-varying coefficients can also be represented in this way.

Moreover, Harvey (1984) has described a general class of trend-and-seasonal models which involve the classical decomposition of a time series into trend, seasonality and irregular variation, but which can also be represented as state-space models. We pay particular attention to these important models. Note that the decomposition must be additive in order to get a linear state-space model. If for example the seasonal effect is thought to be multiplicative, then logarithms must be taken in order to fit a structural model, although this implicitly assumes that the 'error' terms are also multiplicative. A key feature of structural models (and more generally of linear state-space models) is that the observation equation involves a **linear** function of the state variables and yet does not restrict the model to be constant through time. Rather it allows local features, such as trend and seasonality, to be updated through time using the transition equation.

10.1.1 The steady model

Suppose that the observation equation is given by

$$X_t = \mu_t + n_t \qquad (10.4)$$

where the unobservable current level μ_t is assumed to follow a random walk given by

$$\mu_t = \mu_{t-1} + w_t \qquad (10.5)$$

Equation (10.5) is the transition equation. The state vector, θ_t, here consists of a single state variable, namely μ_t and so is a scalar, while h_t and G_t are also constant scalars, namely unity. The model involves two error terms, namely n_t and w_t, which are usually assumed to be independent, normally distributed with zero means and respective variances σ_n^2 and σ_w^2. The ratio of these two variances, namely σ_w^2/σ_n^2, is called the **signal-to-noise ratio** and is an important factor in determining the features of the model. In particular, if $\sigma_w^2 = 0$ then μ_t is a constant and the model reduces to a trivial, constant-mean model.

The state-space model defined by equations (10.4) and (10.5) is customarily called the steady model because there is no trend term included (compare with the linear growth model in Section 10.1.2). The model is very simple but very important since it can be shown that simple exponential smoothing produces optimal forecasts, not only for an ARIMA(0, 1, 1) model (see Chapter 5) but also for this steady model (see Section 10.2 and Exercise 10.1). The reader can readily explore the relation with the ARIMA(0, 1, 1) model by taking first differences of X_t in equation (10.4) and using equation (10.5) to show that the first differences are stationary and have the same autocorrelation function as an MA(1) model. It can also be shown that the steady model and the ARIMA(0, 1, 1) model give rise to the same forecast function.

Thus the steady model could be considered for data showing no long-term trend and no seasonality but some short-term correlation.

10.1.2 The linear growth model

The linear growth model is specified by the three equations

$$X_t = \mu_t + n_t$$

$$\mu_t = \mu_{t-1} + \beta_{t-1} + w_{1,t} \qquad (10.6)$$

$$\beta_t = \beta_{t-1} + w_{2,t}$$

The first equation is the observation equation, while the next two are transition equations. The state vector $\boldsymbol{\theta}_t^T = (\mu_t, \beta_t)$ has two components, which can naturally be interpreted as the local level μ_t and the local trend β_t. Note that the latter state variable does not actually appear in the observation equation, and the reader may readily verify that $h_t^T = (1, 0)$ and

$$G_t = \begin{bmatrix} 1 & 1 \\ 0 & 1 \end{bmatrix}$$

are both constant through time.

The title 'linear growth model' is self-explanatory. The current level μ_t changes linearly through time, but the growth rate (or trend) may also evolve. Of course if $w_{1,t}$ and $w_{2,t}$ have zero variance, then the trend is constant (or deterministic) and we have what is called a **global** linear trend model. However, this situation is unlikely to occur in practice and modern thinking generally prefers a **local** linear trend model where the trend is allowed to change. In any case the global model is a special case of model (10.6) and so it seems more sensible to fit the latter, more general model. It is arguably easier to get a variety of trend models from special cases of a general structural state-space model than from the Box-Jenkins ARIMA class of models.

The reader may easily verify that the second differences of X_t in equation (10.6) are stationary and have the same autocorrelation function as an MA(2) model. In fact it can be shown that two-parameter exponential smoothing (where level and trend are updated) is optimal for an ARIMA(0, 2, 2) model and also for the linear growth model (e.g. see Abraham and Ledolter, 1986).

10.1.3 The basic structural model

There are various ways of incorporating seasonality into a state-space model, such as the following model which is usually called the basic structural model:

$$X_t = \mu_t + i_t + n_t$$

$$\mu_t = \mu_{t-1} + \beta_{t-1} + w_{1,t}$$

$$\beta_t = \beta_{t-1} + w_{2,t} \qquad (10.7)$$

$$i_t = -\sum_{j=1}^{s-1} i_{t-j} + w_{3,t}$$

This model includes a local level μ_t, a local trend β_t, a local seasonal index i_t, and four separate 'error' terms which are all assumed to be additive. If there are s periods in one year (or season), then the fourth equation in (10.7) assumes that the expectation of the sum of the seasonal effects over one year is zero. The state vector now has $s+2$ components, namely $\mu_t, \beta_t, i_t, i_{t-1}, \ldots, i_{t-s+1}$. A full discussion of this model is given by Harvey (1984, 1989) who also discusses various extensions, such as the incorporation of intervention and explanatory variables. Further comments on structural models are given by Chatfield and Yar (1988, Section 6) and Newbold (1988).

10.1.4   State-space representation of an AR(2) process

The AR(2) model can be written as

$$X_t = \phi_1 X_{t-1} + \phi_2 X_{t-2} + Z_t \tag{10.8}$$

Define the (rather artificial) state vector at time t as $\theta_t^T = (X_t, \phi_2 X_{t-1})$. Then the observation equation may be written as

$$X_t = (1, 0)\theta_t$$

with $\sigma_n^2 = 0$, while (10.8) may be written as part of the transition equation

$$\theta_t = \begin{bmatrix} \phi_1 & 1 \\ \phi_2 & 0 \end{bmatrix} \theta_{t-1} + \begin{bmatrix} 1 \\ 0 \end{bmatrix} Z_t \tag{10.9}$$

since $\theta_{t-1}^T = (X_{t-1}, \phi_2 X_{t-2})$.

This looks (and is!) a rather contrived piece of mathematical trickery, and we normally prefer to use equation (10.8) which appears more natural than equation (10.9). (In contrast the state-space linear growth model (10.6) may well appear more natural than an ARIMA(0, 2, 2) model.) However, equation (10.9) does replace two-stage dependence with two equations involving one-stage dependence, and also allows us to use the general results relating to state-space models, such as the recursive estimation of parameters, should we need to do so. For example the Kalman filter provides a general method of estimation for ARIMA models (e.g. Kohn and Ansley, 1986). However, it should also be said that the approach to identifying state-space models is generally quite different to that for ARIMA models.

Note that the state-space representation of an ARMA model is not unique and it may be possible to find many equivalent representations. For example, the reader may like to find an alternative state-space representation of equation (10.8) using the (more natural?) state vector $\theta_t^T = (X_t, X_{t-1})$, or using the (more useful?) state vector $\theta_t^T = [X_t, \hat{X}(t, 1)]$; see Exercise 10.4.

10.1.5   Bayesian forecasting

Bayesian forecasting (Harrison and Stevens, 1976) is a general approach to forecasting which includes a variety of methods, such as regression and exponential smoothing, as special cases. It relies on a model, called the **dynamic linear model**, which is closely related to the general class of state-space models. The Bayesian formulation means that the Kalman filter is regarded as a way of updating the probability distribution of θ_t when a new observation at time t becomes available. The Bayesian approach also enables the analyst to consider the case where several different models are entertained and it is required to choose a single model to represent the process, or alternatively to compute forecasts which are based on several alternative possible models. For example when the latest observation appears to be an outlier, one could entertain the possibility that this represents a step change in the process, or that it arises because of a single intervention, or that it is a 'simple' outlier with no change in the underlying model. The respective probabilities of each model being 'true' are updated after each new observation. An expository introduction is given by Bolstad (1986). The approach has been extended in various ways, for example by considering non-linear and non-normal time series, and the reader is referred to West, Harrison and Mignon (1985).

The approach has some staunch adherents, while others find the avowedly Bayesian approach rather intimidating. This author has no practical experience with the approach, but see Taylor and Thomas (1982) and Fildes (1983). The latter suggests that the method is generally not worth the extra complexity compared with alternative, simpler methods, though there are of course exceptions.

10.1.6   A regression model with time-varying coefficients

Suppose that the observed variable X_t is known to be linearly related to a known explanatory variable u_t by

$$X_t = a_t + b_t u_t + n_t$$

where the regression coefficients a_t and b_t are allowed to evolve through time according to a random walk. If we write $\theta_t^T = [a_t, b_t]$ and $h_t^T = [1, u_t]$, then we may write this model in state-space form by

$$X_t = h_t^T \theta_t + n_t$$
$$\theta_t = \theta_{t-1} + w_t$$

(10.10)

Of course if the elements of w_t have zero variance, then θ_t is constant and we are back to the familiar linear regression model with constant coefficients. The advantage of equation (10.10) is that we can consider a much more general class of models, which includes simple regression as a special case, and then apply the general theory relating to state-space models.

10.1.7 Model building

An important difference between state-space modelling in time-series applications and in some engineering problems is that the structure and properties of a time series may not be known *a priori*. In order to apply state-space theory, we need to know $\mathbf{h}_t$ and G_t in the model equations and also to know the variances and covariances of the disturbance terms, namely σ_n^2 and W_t. The choice of $\mathbf{h}_t$ and G_t (i.e. the choice of a suitable state-space model) may be accomplished using a variety of aids including external knowledge and a preliminary examination of the data. For example Harvey (1984) claims that the basic structural model (see Section 10.1.3) can describe many time series with trend and seasonal terms, but the analyst must for example check that the seasonal variation is additive (or consider a transformation). In other words the use of a state-space model does not take away the usual problem of finding a suitable type of model. (Model fitting is usually easy, but model building can be hard.)

The other problem in time-series applications is that the error variances are usually unknown. This can be overcome by guesstimating them, by updating them in an appropriate way, or, as recommended by Harvey (1984), by estimating them from a set of data over a suitable fit period.

10.2 THE KALMAN FILTER

In state-space modelling, the prime objective is to estimate the signal in the presence of noise. In other words we want to estimate the state vector θ_t. The Kalman filter provides a general method of doing this. It consists of a set of equations which allows us to update the estimate of θ_t when a new observation becomes available. This updating procedure has two stages, called the prediction stage and the updating stage.

Suppose we have observed a time series up to time $t-1$, and that $\hat{\theta}_{t-1}$ is the 'best' estimator for θ_{t-1} based on information up to this time. By 'best' we mean that it is the minimum mean square error estimator. Further, suppose that we have evaluated the variance-covariance matrix of $\hat{\theta}_{t-1}$ which we denote by P_{t-1}. The first stage, called the **prediction** stage, is concerned with forecasting θ_t from time $t-1$, and we denote the resulting estimator in an obvious notation by $\hat{\theta}_{t|t-1}$. Considering equation (10.3), where $\mathbf{w}_t$ is still unknown at time $t-1$, the obvious estimator for θ_t is given by

$$\hat{\theta}_{t|t-1} = G_t \hat{\theta}_{t-1} \qquad (10.11)$$

with variance-covariance matrix

$$P_{t|t-1} = G_t P_{t-1} G_t^T + W_t \qquad (10.12)$$

Equations (10.11) and (10.12) are the prediction equations. Equation (10.12) follows from standard results on variance-covariance matrices for vector random variables (e.g. Chatfield and Collins, 1980, equation (2.9)).

When the new observation at time t, X_t, becomes available, the estimator of $\boldsymbol{\theta}_t$ can be modified to take account of this extra information. The prediction error is given by

$$e_t = X_t - \mathbf{h}_t^\mathsf{T} \hat{\boldsymbol{\theta}}_{t|t-1}$$

and it can be shown that the updating equations are given by

$$\hat{\boldsymbol{\theta}}_t = \hat{\boldsymbol{\theta}}_{t|t-1} + K_t e_t \qquad\qquad (10.13)$$

and

$$P_t = P_{t|t-1} - K_t \mathbf{h}_t^\mathsf{T} P_{t|t-1} \qquad\qquad (10.14)$$

where

$$K_t = P_{t|t-1} \mathbf{h}_t / [\mathbf{h}_t^\mathsf{T} P_{t|t-1} \mathbf{h}_t + \sigma_n^2] \qquad\qquad (10.15)$$

is called the Kalman gain matrix, which in the univariate case is just a vector. Equations (10.13) and (10.14) constitute the second stage of the Kalman filter and are called the **updating** equations.

We will not attempt to derive the updating equations or to demonstrate the optimality of the Kalman filter. However, we note that the results may be found via least squares theory or using a Bayesian approach. A clear introduction to the Kalman filter is given by Meinhold and Singpurwalla (1983), while more detailed accounts are given by Abraham and Ledolter (1983, Section 8.3.1), Harvey (1981a, Chapter 4) and Aoki (1987).

A major practical advantage of the Kalman filter is that the calculations are recursive, so that although the current estimates are based on the whole past history of measurements, there is no need for an ever-expanding memory. Recursive methods, such as exponential smoothing, are increasingly popular in many areas of statistics. A second advantage of the Kalman filter is that it converges fairly quickly when there is a constant underlying model, but can also follow the movement of a system where the underlying model is evolving through time.

The Kalman filter equations look rather complicated at first sight, but they may readily be programmed in their general form and reduce to much simpler equations in certain special cases. For example, consider the steady model of Section 10.1.1 where the state vector $\boldsymbol{\theta}_t$ consists of just one state variable, the current level μ_t. After some algebra (e.g. Abraham and Ledolter, 1986), it can be shown that the Kalman filter for this model in the steady-state case (as $t \to \infty$) reduces to the simple recurrence relation

$$\hat{\mu}_t = \hat{\mu}_{t-1} + \alpha e_t \qquad\qquad (10.16)$$

where the smoothing constant α is a (complicated) function of the signal-to-noise ratio σ_w^2/σ_n^2 (see Exercise 10.1). When this ratio tends to zero, so that μ_t is a constant, we find that α tends to zero as would intuitively be expected, while as σ_w^2/σ_n^2 becomes large, then α approaches unity. Equation (10.16) is of course simple exponential smoothing.

As a second example, consider the linear regression model with time-varying coefficients in Section 10.1.6. Abraham and Ledolter (1983, Section 8.3.3) show how to find the Kalman filter for this model. In particular it is easy to demonstrate that, when W_t is the zero matrix, so that the regression coefficients are constant, then G_t is the identity matrix, $P_{t|t-1} = P_{t-1}$, and the Kalman filter reduces to the equations

$$\boldsymbol{\theta}_t = \boldsymbol{\theta}_{t-1} + K_t e_t$$

$$P_t = P_{t-1} - K_t \mathbf{h}_t^T P_{t-1}$$

where

$$e_t = X_t - \mathbf{h}_t^T \boldsymbol{\theta}_{t-1}$$

$$K_t = P_{t-1} \mathbf{h}_t [\mathbf{h}_t^T P_{t-1} \mathbf{h}_t + \sigma_n^2]^{-1}$$

Abraham and Ledolter (1983, Section 8.3.3) demonstrate that these equations are the same as the 'well-known' updating equations for recursive least squares provided that starting values are chosen in an appropriate way.

In order to initialize the Kalman filter, we need values for $\boldsymbol{\theta}_t$ and P_t at the start of the series. This can be done by *a priori* guesswork, relying on the fact that the Kalman filter will rapidly update these quantities so that the initial choices become dominated by the data. Alternatively, one may be able to estimate the $(m \times 1)$ vector $\boldsymbol{\theta}_t$ at time $t = m$ by least squares from the first m observations, since if we can write

$$\mathbf{X}_0 = M\boldsymbol{\theta}_m + \mathbf{e}$$

where $\mathbf{X}_0^T = (X_m, X_{m-1}, \ldots, X_1)$, M is a known non-singular $(m \times m)$ matrix, and $\mathbf{e}$ is an m-vector of independent 'error' terms, then

$$\hat{\boldsymbol{\theta}}_m = M^{-1} \mathbf{X}_0 \tag{10.17}$$

is the least squares estimate of $\boldsymbol{\theta}_m$ (since M is a square matrix). An example is given in Exercise 10.2.

Forecasts may easily be obtained from the state-space model. At time t, the k-step-ahead forecast is given by

$$\hat{X}(t, k) = \mathbf{h}_{t+k}^T \hat{\boldsymbol{\theta}}_{t+k}$$

$$= \mathbf{h}_{t+k}^T G_{t+k} G_{t+k-1} \cdots G_{t+1} \hat{\boldsymbol{\theta}}_t$$

Of course if G_t is a constant, say G, then

$$\hat{X}(t, k) = \mathbf{h}_{t+k}^T G^k \hat{\boldsymbol{\theta}}_t \tag{10.18}$$

The Kalman filter is applied to state-space models which are linear in the parameters. In practice many time-series models, such as multiplicative seasonal models, are non-linear. Then it may be possible to apply a filter, called the **extended Kalman filter**, by making a locally linear approximation to the model. Applications to data where the noise is not necessarily normally distributed are also possible (Kitagawa, 1987), but we will not pursue these more advanced topics here.

10.2.1   The linear growth model

We will evaluate the Kalman filter for the linear growth model of Section 10.1.2. Suppose that from data up to time $(t-1)$ we have estimates $\hat{\mu}_{t-1}$ and $\hat{\beta}_{t-1}$ of the level and trend. At time $(t-1)$ the best forecasts of $w_{1,t}$ and $w_{2,t}$ are both zero so that the best forecasts of μ_t and β_t in equation 10.6 are clearly given by

$$\hat{\mu}_{t|t-1} = \hat{\mu}_{t-1} + \hat{\beta}_{t-1}$$

and

$$\hat{\beta}_{t|t-1} = \hat{\beta}_{t-1}$$

These agree with equation (10.11). When X_t becomes available, we can find $e_t = X_t - \hat{\mu}_{t|t-1}$ so that we may use equation (10.13) to give

$$\hat{\mu}_t = \hat{\mu}_{t|t-1} + C_{1,t}e_t = \hat{\mu}_{t-1} + \hat{\beta}_{t-1} + C_{1,t}e_t$$

and

$$\hat{\beta}_t = \hat{\beta}_{t|t-1} + C_{2,t}e_t = \hat{\beta}_{t-1} + C_{2,t}e_t$$

where $C_{1,t}$, $C_{2,t}$ are the elements of the Kalman gain 'matrix' (here a 2×1 vector), K_t, which can be evaluated after some algebra. It is interesting to note that these two equations are of similar form to those in the 2-parameter non-seasonal version of Holt-Winters (see Section 5.2.3). There the level and trend are denoted by L_t, T_t respectively and we have for example that

$$L_t = \alpha X_t + (1-\alpha)(L_{t-1} + T_{t-1})$$
$$= L_{t-1} + T_{t-1} + \alpha e_t$$

where $e_t = X_t - [L_{t-1} + T_{t-1}]$. In the steady state as $t \to \infty$, $C_{1,t}$ tends to a constant which corresponds to the smoothing parameter α. This demonstrates that 2-parameter Holt-Winters is optimal for the linear growth model.

An intuitively obvious way to initialize the two state variables from the first two observations is to take $\hat{\mu}_2 = X_2$ and $\hat{\beta}_2 = X_2 - X_1$ (see Exercise 10.2).

EXERCISES

10.1 Consider the steady model in Section 10.1.1, and denote the signal-to-noise ratio σ_w^2/σ_n^2 by c. Show that the first-order autocorrelation coefficient of $(1-B)X_t$ is $-1/(2+c)$ and that higher-order autocorrelations are all zero.

For the ARIMA(0, 1, 1) model

$$(1-B)X_t = Z_t + \theta Z_{t-1}$$

show that the first-order autocorrelation coefficient of $(1-B)X_t$ is $\theta/(1+\theta^2)$ and that higher-order autocorrelations are all zero. Thus the two models have equivalent autocorrelation properties when $\theta/(1+\theta^2) = -1/(2+c)$. Hence show that the invertible solution with $|\theta| < 1$ is $\theta = \frac{1}{2}[(c^2+4c)^{1/2} - c] - 1$.

Applying the Kalman filter to the steady model we find, after some algebra, that in the steady state (as $t\to\infty$ and $P_t\to$constant) we have

$$\hat{\mu}_t = \hat{\mu}_{t-1} + \alpha e_t$$

and

$$\alpha = 1 + \theta = \frac{1}{2}[(c^2+4c)^{1/2} - c]$$

and this is simple exponential smoothing. Now the ARIMA model is invertible provided that $-1 < \theta < 1$, suggesting that $0 < \alpha < 2$. However, the steady model restricts α to the range $0 < \alpha < 1$ (and hence $-1 < \theta < 0$) and physical considerations suggest that this is generally a more sensible model. Do you agree?

10.2 Consider the following special case of the linear growth model:

$$X_t = \mu_t + n_t$$
$$\mu_t = \mu_{t-1} + \beta_{t-1}$$
$$\beta_t = \beta_{t-1} + w_t$$

where n_t, w_t are independent normal with zero means and respective variances σ_n^2, σ_w^2. Show that the initial least squares estimator of the state vector at time $t=2$, in terms of the observations X_1 and X_2, is $[\hat{\mu}_2, \hat{\beta}_2] = [X_2, X_2 - X_1]$ with variance-covariance matrix

$$P_2 = \begin{bmatrix} \sigma_n^2 & \sigma_n^2 \\ \sigma_n^2 & 2\sigma_n^2 + \sigma_w^2 \end{bmatrix}$$

If $\sigma_w^2 = 0$, so that we have ordinary linear regression with constant coefficients, and a third observation X_3 becomes available, apply the Kalman filter to show that the estimator of the state vector at time $t=3$ is given by

$$[\hat{\mu}_3, \hat{\beta}_3] = [\tfrac{5}{6}X_3 + \tfrac{1}{3}X_2 - \tfrac{1}{6}X_1, (X_3 - X_1)/2]$$

Verify that these are the same results that would be obtained by ordinary least squares regression.

10.3   Find a state-space representation of (a) the MA(1) process $X_t = Z_t + \beta Z_{t-1}$, (b) the MA(2) process.
(Hint for (a): Try $\boldsymbol{\theta}_t^T = [X_t, \hat{X}(t, 1)] = [X_t, \beta Z_t]$.)

10.4   Find a state-space representation of the AR(2) process in equation (10.8) based on the state vector $\boldsymbol{\theta}_t^T = (X_t, X_{t-1})$, and show that

$$G = \begin{bmatrix} \phi_1 & \phi_2 \\ 1 & 0 \end{bmatrix}$$

Also find the state-space representation based on the state vector $\boldsymbol{\theta}_t^T = [X_t, \hat{X}(t, 1)]$, where $\hat{X}(t, 1)$ is the optimal one-step-ahead predictor at time t, namely $\phi_1 X_t + \phi_2 X_{t-1}$, and show that

$$G = \begin{bmatrix} 0 & 1 \\ \phi_2 & \phi_1 \end{bmatrix}$$

with $\mathbf{w}_t^T = (1, \phi_1)Z_t$.

11
Some other topics

This chapter provides a brief introduction to a number of topics not covered earlier in this book. References are provided to enable the reader to get further details if desired. Further up-to-date reviews of many standard and non-standard aspects of time-series analysis, including recent research developments, are given by Cox (1981), Newbold (1981, 1984, 1988) and in the collections of papers edited by Brillinger and Tiao (1980), Brillinger and Krishnaiah (1983) and Hannan, Krishnaiah and Rao (1985).

11.1 CONTROL THEORY

To many statisticians, the word 'control' implies statistical quality control using control charts, but in this section we use the word as in control engineering to denote the search for an automatic control procedure for a system whose structure may or may not be known. This is often the ultimate objective in identifying linear systems as in Chapter 9.

There are many different approaches to control, such as those based on linear parametric models of the form (cf. (9.41))

$$\delta(B)Y_t = \omega(B)X_t + \theta(B)Z_t$$

where $\{Y_t\}$ denotes an 'output', $\{X_t\}$ denotes an 'input', $\{Z_t\}$ denotes a purely random process, and δ, ω, θ are polynomials in the backward shift operator B. Many other approaches have been described, including the use of cross-spectral analysis and stochastic adaptive systems (e.g. Harris, 1976, Section 7), but we do not have space to discuss them all here. The main contributions to control theory have naturally been made by control engineers, though mathematicians, statisticians and operational researchers have also played a part.

Control engineers were originally concerned mainly with deterministic control (and many of them still are). Some of the reported research, such as the solution of non-linear differential equations subject to boundary conditions, has an obvious relevance to control theory but may equally be regarded as a branch of applied mathematics. In recent years attention has widened from deterministic problems to the consideration of stochastic control, where the system being controlled is subject to random disturbances. Some references covering the whole field include Astrom (1970), Fuller (1970), Jacobs (1974), Harris (1976), Priestley (1981) and Davis and Vinter (1985).

In stochastic control, a basic problem is that of separating the signal from the noise (see Section 5.6), and there has been much work on the filtering problem starting with the work of Wiener and Kolmogorov. One major development has been the Kalman filter, which is a recursive method of estimating the state of a system in the presence of noise (see Chapter 10). Kalman filtering has been used in many applications including the control of a space rocket, where the system dynamics are well defined but the disturbances are unknown.

Although many statistical problems arise in control theory, statisticians have made a relatively small contribution to the subject. Box and Jenkins (1970) are one exception, and these authors show how to identify a linear parametric model for a system (see Section 9.4.2) and hence find a 'good' control procedure. The Box-Jenkins approach has been compared with the control engineer's state-space representation approach by Priestley (1981). It is regrettable that there has been relatively little communication between statisticians and control engineers, partly because of understandable differences in emphasis and interest. In particular, Box and Jenkins (1970) stress the need to identify systems from observed data, while much of the control literature (e.g. Nicholson, 1980) appears to assume knowledge of the system structure. It is to be hoped that the future will bring more collaboration.

11.2 MODELLING NON-STATIONARY SERIES

Much of the theory in the time-series literature is applicable to stationary processes. In practice most real time series do change with time, albeit slowly in many cases.

It is important to understand the many different types of non-stationarity which may arise and methods for dealing with them (see also Cox, 1981, Section 4; Priestley, 1988, Chapter 6). For example, we have already discussed various methods of transforming data to stationarity (e.g. differencing) in order to be able to fit stationary models and use the theory of stationary processes. This approach is ideal when the non-stationary features are not of primary concern, as for example when instrument drift arises. Alternatively the non-stationary features of the data may be of intrinsic interest in themselves, in which case it may be more rewarding to model them explicitly, rather than remove them and concentrate on modelling the stationary residuals. Thus there is, for example, a fundamental difference between fitting an ARIMA model, which describes non-stationary features implicitly, and fitting a state-space model, which describes them explicitly.

Slow changes in mean are one common source of non-stationarity. If a global (deterministic) function of time can be assumed, then such components can easily be fitted and removed (e.g. by fitting a polynomial). However, it is now more common to assume that there are local changes and perhaps fit a local linear trend which is updated through time. Some sort of filtering or differencing may then be employed, the choice depending in part on whether the differenced series has a natural interpretation. Cyclical changes in mean (e.g. seasonality) can also be dealt with by filtering, by differencing or by fitting a global model such as a few sine and cosine terms of appropriate frequency.

If we consider a linear model, such as an AR process, then it is useful to distinguish several different ways in which non-stationarity may arise. If the coefficients do not satisfy the stationarity conditions, then the roots of equation (3.5) may lie on the unit circle, in which case the series can be made stationary by differencing. Alternatively, the roots may lie outside the unit circle, leading to explosive behaviour. Another possibility is that the coefficients are changing through time, perhaps suddenly (e.g. Tyssedal and Tjostheim, 1988) or perhaps slowly, the latter being one example of a slow change in the underlying model structure. The latter can also be studied in the frequency domain, and there are ways of generalizing the spectrum to cope with non-stationary behaviour. The use of **evolutionary spectra** (Priestley, 1981, Chapter 11; 1988) is one possibility. **Complex demodulation** is an alternative approach, which studies signals in a narrow frequency band to see how they change through time (e.g. Bloomfield, 1976; Hasan, 1983). More generally, even with a long, apparently stationary series, it is still a good idea to split the series into reasonably long, non-overlapping segments and compare the properties of the segments, particularly the general form of the ac.f. and spectrum.

Finally, the most difficult type of non-stationarity to handle is a sudden change in structure, due perhaps to the occurrence of a known external event such as an earthquake or a labour dispute. The change may produce a short-term transient effect or a long-term change in the model structure. One or more outliers may be visible in the time plot and these can create problems with standard methods. For example, Hillmer (1984) shows how one outlier can affect several consecutive forecasts unless adjusted, while Chang, Tiao and Chen (1988) discuss parameter estimation in the presence of outliers. Box and Tiao (1975) show how to model sudden changes with a technique called **intervention analysis**, which is somewhat similar to the use of dummy variables in regression. Alternatively, **Bayesian forecasting** (see Chapter 10) can allow one to specify a range of models which can deal with outliers and step changes in the mean and trend. However, in some situations it is wiser to accept that there may be no sensible way to model a process. For example, Figure 5.1(a) shows data on the sales of insurance policies where I was asked to produce forecasts. It is clear that there has been a sudden change in the underlying structure. Rather than try to model this, it is more important to ask questions to get appropriate background information. In this case I found that the two peaks corresponded to two sales drives. Thus the most important step is to ask if there is going to be a third sales drive. In the absence of such information, it would be most unwise to try and produce forecasts.

11.3 NON-LINEAR MODELS

Much of the time-series literature not only assumes that all series are stationary (or can be reduced to stationarity by a simple transformation such as differencing) but also assumes that the appropriate model will be **linear**, so that the observed series can be represented as a linear function of the present and past values of a purely random process. For example the AR, MA, ARMA and ARIMA models are all linear (see Section 3.4.7 for a discussion of the general linear process).

However, there is no reason why real generating processes should all be linear, and various types of non-linear model have been studied in recent years (e.g. Priestley, 1981, Chapter 11; 1988; Newbold, 1984; Granger and Newbold, 1986, Chapter 10). One class of non-linear models, called the bilinear class, may be regarded as the natural non-linear extension of an ARMA model. For example the first-order bilinear model is given by

$$X_t = aX_{t-1} + bZ_t + cZ_{t-1}X_{t-1}$$

where $\{Z_t\}$ denotes a purely random process and a, b, c are parameters. It is the last term on the right-hand side of the above equation which is the non-linear term.

Another class of models are threshold autoregressive (TAR) models where the model parameters depend on the past values of the process. For example, a first-order TAR model could be

$$X_t = \begin{cases} \alpha_1 X_{t-1} + Z_t & \text{if } X_{t-1} < d \\ \alpha_2 X_{t-1} + Z_t & \text{if } X_{t-1} \geqslant d \end{cases}$$

where the constant d is called the threshold.

One problem with non-linear models is that it may be difficult to evaluate conditional expectations more than one step ahead, and so forecasts may be difficult to find. Also note that it is sometimes the case that the width of prediction intervals does not increase with the lead time (e.g. if the model gives rise to what are called limit cycles); or alternatively that a sensible prediction interval may comprise two disjoint intervals, and the conditional expectation by itself may be misleading.

Note that the term 'non-linear' refers to the form of the relationship. An MA model, for example, is a linear process but may be regarded as non-linear in the parameters in that the one-step-ahead errors are non-linear functions of the parameters, so that explicit estimators may not be available (see Section 4.3.1).

11.4 MODEL IDENTIFICATION TOOLS

The two standard tools which are used to identify an appropriate ARMA model for a given stationary time series are the sample ac.f. and the sample partial ac.f. (see Chapter 4). Recently several new tools have been proposed, and a detailed review of methods for determining the order of an ARMA process is given by de Gooijer *et al.* (1985); see also Newbold (1988).

An alternative to the partial ac.f. is the inverse ac.f., whose use in identifying ARMA models is described by Chatfield (1979). The inverse ac.f. of the ARMA model (see equation (3.6a))

$$\phi(B)X_t = \theta(B)Z_t$$

is exactly the same as the ordinary ac.f. of the corresponding inverse ARMA model given by

$$\theta(B)X_t = \phi(B)Z_t$$

where ϕ and θ are interchanged. It turns out that the inverse ac.f. has similar properties to the partial ac.f. in that it 'cuts off' at lag p for an AR(p) process but generally dies out slowly for MA and ARMA processes. In other words it has inverse properties to the ordinary ac.f., which 'cuts off' for MA processes. Chatfield (1979) argues that the inverse ac.f. often contains more information than the partial ac.f., extends easily to seasonal ARIMA models, and may well displace the partial ac.f.

Instead of subjectively examining functions like the ac.f., an alternative type of approach is to choose the ARMA model which optimizes a suitably chosen function of the data. One approach, based on Akaike's final prediction error (FPE) criterion, is concerned with comparing AR processes of different order. The order p is essentially selected so as to get the estimated one-step-ahead predictor with the smallest mean square error (Priestley, 1981, p. 372). A more general criterion for model selection is to minimize a quantity called Akaike's information criterion (AIC), which is equal to $[-2 \ln \text{ (maximized likelihood)} + 2 \text{ (number of independent parameters estimated)}]$. The AIC criterion can be used to compare ARMA models as well as AR models. A third criterion is Parzen's autoregressive transfer function criterion (CAT), which also tends to give similar results (e.g. Landers and Lacoss, 1977). Unfortunately all these criteria may give more than one minimum, depend on assuming that the data are normally distributed, and tend to identify too many parameters. Akaike has recently developed a Bayesian modification of AIC, denoted by BIC, which penalizes models with large numbers of parameters. If the number of independent parameters is denoted by p, and N denotes the number of observations to which the model is fitted, then BIC replaces the term $2p$ in the AIC criterion by $(p + p \log N)$. The Schwartz criterion is yet another alternative, which is somewhat similar to BIC in its dependence on $\log N$. Priestley (1981, Chapter 5) gives a general review. It is clear that these criteria should be used only as guides, and many computer packages routinely print several of them to compare.

11.5   AUTOREGRESSIVE SPECTRUM ESTIMATION

Spectral analysis is concerned with estimating the spectrum of a stationary stochastic process. The approach described in Chapter 7, which is based on Fourier analysis, is essentially non-parametric in that no model is assumed *a priori*. In recent years a parametric approach called autoregressive spectrum estimation has become a popular alternative method, and this will now be briefly described.

Many stationary stochastic processes can be adequately approximated by an AR process of sufficiently high order, say

$$X_t = \alpha_0 + \alpha_1 X_{t-1} + \cdots + \alpha_p X_{t-p} + Z_t \tag{11.1}$$

The spectrum of (11.1) at frequency ω is inversely proportional to

$$\left| 1 - \sum_{k=1}^{p} \alpha_k e^{-ik\omega} \right|^2$$

The procedure requires the user to assess the order p of the process, possibly using Akaike's information criterion. Then the AR parameters are estimated, as described in Section 4.2, and substituted into the expression for the spectrum of (11.1). The approach tends to give a smoother spectrum than that given by the non-parametric approach, though it can also pick out fairly narrow peaks in the spectrum. Convincing examples are given by Jones (1974), Griffiths and Prieto-Diaz (1977) and Tong (1977), and the approach is certainly worth considering. Note that the selected order p of the process may be higher than might be expected, as values around 10 are common and values as high as 28 have been reported.

An obvious development of this approach is to fit ARMA models rather than AR models to give what might be called ARMA spectrum estimation, but the simplicity of fitting AR models is likely to inhibit such a development.

11.6   SPATIAL SERIES

In some scientific areas, particularly in ecology and agriculture, data often arise which are ordered with respect to one or two **spatial** co-ordinates rather than with respect to time. Methods of analysing spatial data have been described, for example, by Besag (1974) and Cliff and Ord (1975). Although there are many similarities between time-series and spatial analysis, there are also fundamental differences (Chatfield, 1977, Section 9). Ripley (1981) provides comprehensive coverage of spatial methods.

11.7 'CROSSING' PROBLEMS

If one draws a horizontal line (an axis) through a stationary time series, then the time series will cross and recross the line in alternate directions. The times at which the series cuts the axis provide valuable information about the properties of the series. Indeed in some practical problems the crossing points are the only information available. The problems of inferring the properties of a time series from its 'level-crossing' properties are discussed by Cramer and Leadbetter (1967), and a comprehensive survey and bibliography is given by Blake and Lindsey (1973).

11.8 OBSERVATIONS AT UNEQUAL INTERVALS

This book has been mainly concerned with discrete time series measured at equal intervals of time. When observations are taken at unequal intervals, either by accident or design, the general definition of the periodogram (equation (7.17)) still applies but the values of t are no longer integers. Autocorrelation coefficients can only be obtained directly by fitting a smoothed approximation to the time series. **Splines**, which are piecewise polynomials, can be used to provide such an approximation (e.g. see Wegman and Wright, 1983, Section 4). Further references on time series with unequally spaced or missing observations are given by Priestley (1981, p. 586) and Hannan (1970, p. 48). Additional references include Roberts and Gaster (1980), Jones (1985) and the collection of papers edited by Parzen (1984).

11.9 MODELLING MULTIVARIATE TIME SERIES

Observations are often taken simultaneously on two or more time series. For example, in meteorology we might observe temperature, air pressure, rainfall etc. at the same site for the same sequence of time points. Then it may be of interest to develop a multivariate model to describe the interrelationships among the series, and there is growing interest in this topic arising partly from improvements in computing capabilities.

It is helpful to distinguish between the case where the variables arise 'on an equal footing', and the case where one wants to 'explain' the variation in one or more response variables in terms of the variation in other predictor, or explanatory, variables. Regression models (see Section 5.3) are of the latter type, as are transfer-function models (sec Section 9.4). Here we concentrate on multivariate or vector ARMA, models which will be abbreviated as VARMA models. VARMA models are a natural extension of the univariate ARMA models in equation (3.6a) when written in the form

$$\phi(B)\mathbf{X}_t = \theta(B)\mathbf{Z}_t \qquad (11.2)$$

where $\mathbf{X}_t$ is a $(k \times 1)$ vector of observed variables, $\mathbf{Z}_t$ is a $(k \times 1)$ vector of 'noise' variables, and ϕ, θ are matrix polynomials in the backward shift operator B of order p, q respectively, such that

$$\phi(B) = I - \phi_1 B - \cdots - \phi_p B^p$$

$$\theta(B) = I - \theta_1 B - \cdots - \theta_q B^q$$

where I is the $(k \times k)$ identity matrix and $\{\phi_i\}$, $\{\theta_i\}$ are $(k \times k)$ matrices of parameters. If $\phi(B)$ includes a factor of the form $I(1 - B)$, then the model acts on the first differences of the components of $\mathbf{X}_t$ and we have what is called a VARIMA model (although it may not be optimal in practice to difference each component of $\mathbf{X}_t$ in the same way) One problem with VARMA (or VARIMA) models is that there may be different but equivalent (or exchangeable) ways of writing the same model, and there are various ways of imposing constraints on the parameters of (11.2) to ensure uniqueness of representation.

Note that (11.2) reduces to the familiar univariate ARMA model when $k = 1$. Also note that transfer-function models may be regarded as special cases of VARMA models. If the first component of $\mathbf{X}_t$ is the output (or response variable) in an open-loop system, then the parameter matrices in ϕ and θ are upper triangular. In contrast, in a closed-loop system, the 'outputs' feed back to affect the 'inputs' and the general VARMA model may be appropriate to describe the behaviour of the mutually dependent variables. Finally, we note that VARMA models can be further generalized by adding terms involving additional exogenous variables to the right-hand side of equation (11.2), and such a model is sometimes abbreviated as a VARMAX model.

There are various approaches to the identification of VARMA models, which involves assessing the orders p and q, estimating the parameter matrices of ϕ and θ and estimating the variance-covariance matrix of the 'noise' components. Some recent references include Priestley (1981), Tiao and Box (1981), Newbold (1984), Granger and Newbold (1986) and Pena and Box (1987), and there is much current research. Identification is inevitably a complicated process because the number of model parameters increases quadratically with k and can become uncomfortably large unless some constraints are placed on the model. It can be helpful to use external knowledge or a preliminary analysis to identify coefficient matrices where most of the parameters can *a priori* be taken to be zero. Such matrices are called **sparse** matrices. Alternatively, it may be fruitful to use vector AR (VAR) models as a, hopefully adequate, approximation to VARMA models, although there is still a danger of overfitting, and VAR models may not provide as parsimonious an approximation as in the univariate case. Bayesian vector autoregression essentially aims to shrink parameters higher than first order towards zero. Alternatively, with k large, a completely different type of approach is to use some sort of multivariate transformation to reduce the effective dimensionality of the data.

One important tool in VARMA model identification is the matrix of cross-correlation coefficients. The case of two time series has already been partially considered in Chapters 8 and 9, although even here the author must admit that he has typically found it difficult to interpret cross-correlation (and cross-spectral) estimates. The analysis of three or more series is in theory a natural extension, but in practice is much more difficult and should only be attempted by analysts with experience in univariate ARIMA model building. As previously noted, the interpretation of cross-correlations is complicated by the possible presence of autocorrelation within the individual series and by the possible presence of feedback between the series, and one unresolved question is the extent to which series should be filtered or prewhitened before looking at cross-correlations.

Appendix A
The Fourier, Laplace and Z transforms

This appendix provides a brief introduction to the Fourier transform, which is a valuable mathematical tool in time-series analysis. The related Laplace and Z transforms are also introduced.

Given a (possibly complex-valued) function $h(t)$ of a real variable t, the Fourier transform of $h(t)$ is usually defined as

$$H(\omega) = \int_{-\infty}^{\infty} h(t) e^{-i\omega t} \, dt \tag{A.1}$$

provided the integral exists for every real ω. Note that $H(\omega)$ is in general complex. A sufficient condition for $H(\omega)$ to exist is

$$\int_{-\infty}^{\infty} |h(t)| \, dt < \infty$$

If (A.1) is regarded as an integral equation for $h(t)$ given $H(\omega)$, then a simple inversion formula exists of the form

$$h(t) = \frac{1}{2\pi} \int_{-\infty}^{\infty} H(\omega) e^{i\omega t} \, d\omega \tag{A.2}$$

and $h(t)$ is called the inverse Fourier transform of $H(\omega)$, or sometimes just the Fourier transform of $H(\omega)$. The two functions $h(t)$ and $H(\omega)$ are commonly called a Fourier transform pair.

The reader is warned that many authors use a slightly different definition of a Fourier transform to (A.1). For example, some authors put a constant $1/\sqrt{(2\pi)}$ outside the integral in (A.1) and then the inversion formula for $h(t)$ is symmetric. In time-series analysis many authors (e.g. Cox and Miller, 1968, p. 315) put a constant $1/2\pi$ outside the integral in (A.1). The inversion formula then has a unity constant outside the integral.

Some authors in time-series analysis (e.g. Jenkins and Watts, 1968) define

Fourier transforms in terms of the variable $f = \omega/2\pi$ rather than ω. We then find that the Fourier transform pair is

$$G(f) = \int_{-\infty}^{\infty} h(t)e^{-2\pi ift}\,dt \tag{A.3}$$

$$h(t) = \int_{-\infty}^{\infty} G(f)e^{2\pi ift}\,df \tag{A.4}$$

Note that the constant outside each integral is now unity.

In time-series analysis, we will often use the discrete form of the Fourier transform when $h(t)$ is only defined for integer values of t. Then

$$H(\omega) = \sum_{t=-\infty}^{\infty} h(t)e^{-i\omega t} \qquad -\pi \leqslant \omega \leqslant \pi \tag{A.5}$$

is the discrete Fourier transform of $h(t)$. Note that $H(\omega)$ is only defined in the interval $[-\pi, \pi]$. The inverse transform is

$$h(t) = \frac{1}{2\pi} \int_{-\pi}^{\pi} H(\omega)e^{i\omega t}\,d\omega \tag{A.6}$$

Fourier transforms have many useful properties, some of which are used during the later chapters of this book. However, we do not atempt to review them all here. The reader is referred for example to Hsu (1967).

One special type of Fourier transform arises when $h(t)$ is a real-valued even function such that $h(t) = h(-t)$. The autocorrelation function of a stationary time series has these properties. Then, using (A.1) with a constant $1/\pi$ outside the integral, we find

$$H(\omega) = \frac{1}{\pi} \int_{-\infty}^{\infty} h(t)e^{-i\omega t}\,dt$$

$$= \frac{2}{\pi} \int_{0}^{\infty} h(t)\cos \omega t\,dt \tag{A.7}$$

and it is clear that $H(\omega)$ is a real-valued even function. The inversion formula is then

$$h(t) = \frac{1}{2} \int_{-\infty}^{\infty} H(\omega)e^{i\omega t}\,d\omega$$

$$= \int_{0}^{\infty} H(\omega)\cos \omega t\,d\omega \tag{A.8}$$

Equations (A.7) and (A.8) are similar to a Fourier transform pair and are useful when we only wish to define $H(\omega)$ for $\omega > 0$. This pair of equations appears as equations (2.73) and (2.74) in Yaglom (1962). When $h(t)$ is only defined for integer values of t, equations (A.7) and (A.8) become

$$H(\omega) = \frac{1}{\pi} \left\{ h(0) + 2 \sum_{t=1}^{\infty} h(t)\cos \omega t \right\} \tag{A.9}$$

$$h(t) = \int_{0}^{\pi} H(\omega)\cos \omega t \, d\omega \tag{A.10}$$

and $H(\omega)$ is now only defined on $[0, \pi]$.

The Laplace transform of a function $h(t)$ which is defined for $t > 0$ is given by

$$H(s) = \int_{0}^{\infty} h(t)e^{-st} \, dt \tag{A.11}$$

where s is a complex variable. The integral converges when the real part of s exceeds some number called the abscissa of convergence.

The relationship between the Fourier and Laplace transforms is of some interest, particularly as control engineers often prefer to use the Laplace transform when investigating the properties of a linear system, as this will cope with physically realizable systems which are stable **or** unstable. If the function $h(t)$ is such that

$$h(t) = 0 \qquad t < 0 \tag{A.12}$$

and the real part of s is zero, then the Laplace and Fourier transforms of $h(t)$ are the same. The impulse response function of a physically realizable linear system satisfies (A.12) and so for such functions the Fourier transform is a special case of the Laplace transform. More details about the Laplace transform may be found in many books (e.g. Sneddon, 1972).

The Z transform of a function $h(t)$ defined on the non-negative integers is given by

$$H(z) = \sum_{t=0}^{\infty} h(t)z^{-t} \tag{A.13}$$

In discrete time, with a function satisfying (A.12), some authors prefer to use the Z transform rather than the discrete form of the Fourier transform (i.e. (A.5)) or the discrete form of the Laplace transform, namely

$$H(s) = \sum_{t=0}^{\infty} h(t)e^{-st} \tag{A.14}$$

Comparing (A.13) with (A.14) we see that $z = e^{s}$. The reader will observe that when $\{h(t)\}$ is a probability function such that $h(t)$ is the probability of observing the value t, for $t = 0, 1, \ldots$, then (A.13) is related to the probability generating function of the distribution, while (A.5) and (A.14) are related to the moment generating function.

EXERCISES

A.1   If $h(t)$ is real, show that the real and imaginary parts of its Fourier transform, as defined by equation (A.1), are even and odd functions respectively.

A.2   If $h(t)=e^{-a|t|}$ for all real t, where a is a positive real constant, show that its Fourier transform, as defined by equation (A.1), is given by

$$H(\omega)=2a/(a^2+\omega^2) \qquad -\infty<\omega<\infty$$

A.3   Show that the Laplace transform of $h(t)=e^{-at}$ ($t>0$), where a is a real constant, is given by

$$H(s)=1/(s+a) \qquad \text{Re}(s)>a$$

where $\text{Re}(s)$ denotes the real part of s.

Appendix B
The Dirac delta function

Suppose that $\phi(t)$ is any function which is continuous at $t=0$. Then the Dirac delta function $\delta(t)$ is such that

$$\int_{-\infty}^{\infty} \delta(t)\phi(t)\,\mathrm{d}t = \phi(0) \tag{B.1}$$

Because it is defined in terms of its integral properties alone, it is sometimes called the 'spotting' function since it picks out one particular value of $\phi(t)$. It is also sometimes called simply the delta function.

It is important to realize that $\delta(t)$ is **not** a function. Rather it is a generalized function, or distribution, which maps a function into the real line.

Some authors define the delta function by

$$\delta(t) = \begin{cases} 0 & t \neq 0 \\ \infty & t = 0 \end{cases} \tag{B.2}$$

such that

$$\int_{-\infty}^{\infty} \delta(t)\,\mathrm{d}t = 1$$

But while this is often intuitively helpful, it is mathematically meaningless.

The Dirac delta function can also be regarded (e.g. Schwarz and Friedland, 1965) as the limit, as $\varepsilon \to 0$, of a pulse of width ε and height $1/\varepsilon$ (i.e. unit area) defined by

$$u(t) = \begin{cases} 1/\varepsilon & 0 < t < \varepsilon \\ 0 & \text{otherwise} \end{cases}$$

This definition is also not mathematically rigorous, but is heuristically useful. In particular, control engineers can approximate such an impulse by an impulse with unit area whose duration is short compared with the least significant time constant of the response to the linear system being studied.

Even though $\delta(t)$ is a generalized function, it can often be handled as if it were an ordinary function except that we will be interested in the values of integrals involving $\delta(t)$ and never in the value of $\delta(t)$ by itself.

The derivative $\delta'(t)$ of $\delta(t)$ can also be defined by

$$\int_{-\infty}^{\infty} \delta'(t)\phi(t)\,dt = -\phi'(0) \tag{B.3}$$

where $\phi'(0)$ is the derivative of $\phi(t)$ evaluated at $t=0$. The justification for (B.3) depends on integrating by parts as if $\delta'(t)$ and $\delta(t)$ were ordinary functions and using (B.2) to give

$$\int_{-\infty}^{\infty} \delta'(t)\phi(t)\,dt = -\int_{-\infty}^{\infty} \delta(t)\phi'(t)\,dt$$

and then using (B.1). Higher derivatives of $\delta(t)$ may be defined in a similar way.

The delta function has many useful properties (see for example Hsu, 1967).

EXERCISES

B.1   The function $\phi(t)$ is continuous at $t=t_0$. If $a<b$, show that

$$\int_a^b \delta(t-t_0)\phi(t)\,dt = \begin{cases} \phi(t_0) & a<t_0<b \\ 0 & t_0<a, \quad t_0>b \end{cases}$$

B.2   The function $\phi(t)$ is continuous at $t=0$. Show that $\phi(t)\,\delta(t) = \phi(0)\,\delta(t)$.

Appendix C
Covariance

Any reader who is unfamiliar with the laws of probability, distributions, expectation, and basic statistical inference should consult a more elementary text such as Chatfield, C. (1983), *Statistics for Technology* (3rd edn), Chapman and Hall; or Meyer, P. L. (1970), *Introductory Probability and Statistical Applications* (2nd edn), Addison-Wesley.

The idea of covariance is particularly important in the study of time series, and will now be briefly revised.

Suppose two random variables, X and Y, have means $E(X)=\mu_x$, $E(Y)=\mu_y$, respectively. Then the covariance of X and Y is defined to be

$$\text{Cov}(X,\ Y)=E[(X-\mu_x)(Y-\mu_y)]$$

If X and Y are independent, then

$$E[(X-\mu_x)(Y-\mu_y)]=E(X-\mu_x)E(Y-\mu_y)$$
$$=0$$

so that the covariance is zero. If X and Y are **not** independent, then the covariance may be positive or negative depending on whether 'high' values of X tend to go with 'high' or 'low' values of Y.

Covariance is a useful quantity for many mathematical purposes, but it is difficult to interpret as it depends on the units in which X and Y are measured. Thus it is often useful to standardize the covariance between two random variables by dividing by the product of their respective standard deviations to give a quantity called the **correlation** coefficient. It can easily be shown that the correlation coefficient must lie between ± 1 and is a useful measure of the linear association between two variables.

Given N pairs of observations, $\{(x_i, y_i);\ i=1 \text{ to } N\}$, the usual estimator of the covariance between two variables is given by

$$s_{xy}=\sum_{i=1}^{N}(x_i-\bar{x})(y_i-\bar{y})/(N-1)$$

The usual estimate of the correlation coefficient between the two variables is then given by equation (2.2).

If X and Y are random variables from the same stochastic process at different times, then the covariance coefficient is called an **auto**covariance coefficient, and the correlation coefficient is called an **auto**correlation coefficient. If the process is stationary, the standard deviations of X and Y will be the same and their product will be the variance of X (or of Y).

Appendix D
Some worked examples

In this appendix, some detailed worked examples are presented to give more of
the flavour of practical time-series analysis. First I make some general remarks
on the difficulties involved in time-series analysis as well as some comments on
the MINITAB computer package.

D.1 GENERAL COMMENTS

I have been fascinated by time-series analysis since taking a course on the
subject as a postgraduate student. Like most of my fellow students, I found it
difficult but feasible to understand the time-domain material on autocorrelation functions, ARMA models etc., but found spectral analysis initially
incomprehensible. Even today, despite a well-above-average amount of
practice, I still find time-series analysis more difficult than most other
statistical techniques (except perhaps multivariate analysis). This may or may
not be comforting to the reader! It is interesting to assess why difficulties arise,
and this we now do.

The special feature of time-series data is that they are correlated in time
rather than independent. Now most classical statistical inference requires zero
correlation between the observations, and it is generally true that correlated
observations are more difficult to handle. The analysis of time-series data is
often further complicated by the possible presence of trend and seasonal
variation which can be hard to estimate and/or remove. Other practical
problems include missing observations and outliers. But perhaps the major
problem arises from the large subjective element involved in trying to select an
appropriate time-series model for a given set of data. The interpretation of the
correlogram and the sample spectrum is not easy, even after much practice. If
the analysis of a single (linear) time series is not easy, then the analysis of
multivariate, non-stationary and non-linear time series must be even more
difficult, and the reader is not advised to tackle such problems until he has had
plenty of experience with standard univariate problems.

D.2 THE MINITAB COMPUTER PACKAGE

Until recent years, the difficulties of time-series analysis were exacerbated by computing problems. From my point of view, the most welcome recent advance in time-series analysis has been the arrival of the time-series section of a computer package called MINITAB. This easy-to-use interactive package will perform most of the usual statistical tasks such as the calculation of summary statistics, regression and analysis of variance. It is ideal for the student learning statistics, as well as being very useful for the more advanced user. In time series, it will plot data, calculate the autocorrelation function (ac.f.), the partial ac.f., and the cross-correlation function between pairs of series. It will also calculate various differences of the given series and their correlation functions. It will also fit seasonal or non-seasonal ARIMA models and compute forecasts. It does not (at the time of writing) carry out spectral analysis. Further information may be obtained by writing to: Minitab Inc., 3081 Enterprise Drive, State College, PA 16801, USA.

D.3 OTHER COMPUTER PACKAGES

In addition to MINITAB, various other packages and algorithms are available for carrying out various forms of time-series analysis. In fact most general packages offer some time-series options, including BMDP, GEN-STAT and SPSS. The S package will produce very good time plots and also has a variety of time-series modules, including spectral analysis. There are also many more specialized packages such as AUTOBOX for Box-Jenkins modelling, and AUTOCAST and FORECASTPRO for Holt-Winters forecasting. I shall not attempt to review them here because such information dates rapidly. If MINITAB is inadequate or unavailable, the reader is advised to check with his/her computer unit as to what is available.

D.4 EXAMPLES

I have selected three examples to try and give the reader further guidance on practical time-series analysis. In addition to the examples given below there are several more scattered through the text, including three rather brief forecasting problems in Examples 5.1–5.3 and a more detailed spectral analysis in Example 7.1.

Many other examples can be found in the literature, and I note particularly that many univariate and multivariate examples of the Box-Jenkins approach may be found in Jenkins (1979).

Example D.1 Monthly air temperature at Recife   Table D.1 shows the air temperature at Recife in Brazil, in successive months over a 10-year period.

Table D.1   Average monthly air temperatures at Recife for 1953–1962

Year	Jan	Feb	Mar	Apr	May	June	July	Aug	Sept	Oct	Nov	Dec
1953	26.8	27.2	27.1	26.3	25.4	23.9	23.8	23.6	25.3	25.8	26.4	26.9
1954	27.1	27.5	27.4	26.4	24.8	24.3	23.4	23.4	24.6	25.4	25.8	26.7
1955	26.9	26.3	25.7	25.7	24.8	24.0	23.4	23.5	24.8	25.6	26.2	26.5
1956	26.8	26.9	26.7	26.1	26.2	24.7	23.9	23.7	24.7	25.8	26.1	26.5
1957	26.3	27.1	26.2	25.7	25.5	24.9	24.2	24.6	25.5	25.9	26.4	26.9
1958	27.1	27.1	27.4	26.8	25.4	24.8	23.6	23.9	25.0	25.9	26.3	26.6
1959	26.8	27.1	27.4	26.4	25.5	24.7	24.3	24.4	24.8	26.2	26.3	27.0
1960	27.1	27.5	26.2	28.2	27.1	25.4	25.6	24.5	24.7	26.0	26.5	26.8
1961	26.3	26.7	26.6	25.8	25.2	25.1	23.3	23.8	25.2	25.5	26.4	26.7
1962	27.0	27.4	27.0	26.3	25.9	24.6	24.1	24.3	25.2	26.3	26.4	26.7

The individual observations are obtained by averaging over a one-month period.

The objective of our analysis is simply to describe and understand the variation in the data. In fact this set of data has been used throughout the book as an example, and it is convenient to bring the results together here.

The first step is to plot the data, and this has been done in Figure 1.2. The graph shows a large regular seasonal variation with little or no trend. As one would expect, the spectrum of the raw data shows a large peak at a frequency of one cycle per year (see Figure 7.5(a)). The frequency of one cycle per year corresponds to seasonal variation. As the seasonal variation is such a high proportion of the total variation (about 85%), the spectrum tells us very little that is not obvious in the time plot. The correlogram (Figure 2.4(a)) also shows little apart from the seasonal variation.

In order to make any further progress we must remove the seasonal variation from the data. This can be done by seasonal differencing (see Section 4.6) or by simply calculating the overall January average, the overall February average, and so on, and then subtracting each individual value from the appropriate monthly average. We adopted the latter procedure in this case. The correlogram of the seasonally adjusted data is shown in Figure 2.4(b), and this shows that the first three coefficients are significantly different from zero. This indicates some short-term correlation in that a month which is say warmer than average for that month is likely to be followed by one or two further months which are also warmer than their appropriate average.

The periodogram of the seasonally adjusted data is shown in Figure 7.5(c), and exhibits wild fluctuations which are typical of periodograms. Nothing can be learnt from this graph. To estimate the underlying spectrum, we proceed either by smoothing the periodogram or by computing an appropriate weighted transform of the autocorrelation function. The latter approach was used with a Tukey window to produce Figure 7.5(b). The spectrum shows something of a peak at zero frequency, indicating the possible presence of some trend or low-frequency variation, but there is otherwise little to note in the graph.

We summarize our analysis by saying that the series contains high seasonal variation together with a limited short-term correlation effect. The analysis has been rather negative in that we have found little of interest apart from what is evident in the time plot of the data. It is, in fact, a fairly common occurrence in statistics to find that a very simple technique provides most of the information, and this emphasizes the importance of starting any time-series analysis by drawing a time plot.

Example D.2   Yield on short-term government securities   Table D.2 shows the yield on British short-term government securities in successive months over a 21-year period, mainly in the 1950s and 1960s. This series was previously analysed in Chatfield (1978a). The problem is to find a suitable time-series model for the data and compute forecasts for up to 12 months ahead.

Table D.2   Yield on short-term government securities

Year	I	II	III	IV	V	VI	VII	VIII	IX	X	XI	XII
1	2.22	2.23	2.22	2.20	2.09	1.97	2.03	1.98	1.94	1.79	1.74	1.86
2	1.78	1.72	1.79	1.82	1.89	1.99	1.89	1.83	1.71	1.70	1.97	2.21
3	2.36	2.41	2.92	3.15	3.26	3.51	3.48	3.16	3.01	2.97	2.88	2.91
4	3.43	3.29	3.17	3.09	3.02	2.99	2.97	2.94	2.84	2.85	2.86	2.89
5	2.93	2.93	2.87	2.82	2.63	2.33	2.22	2.15	2.28	2.28	2.06	2.54
6	2.29	2.66	3.03	3.17	3.83	3.99	4.11	4.51	4.66	4.37	4.45	4.58
7	4.58	4.76	4.89	4.65	4.51	4.65	4.52	4.52	4.57	4.65	4.74	5.10
8	5.00	4.74	4.79	4.83	4.80	4.83	4.77	4.80	5.38	6.18	6.02	5.91
9	5.66	5.42	5.06	4.70	4.73	4.64	4.62	4.48	4.43	4.33	4.32	4.30
10	4.26	4.02	4.06	4.08	4.09	4.14	4.15	4.20	4.30	4.26	4.15	4.27
11	4.69	4.72	4.92	5.10	5.20	5.56	6.08	6.13	6.09	5.99	5.58	5.59
12	5.42	5.30	5.44	5.32	5.21	5.47	5.96	6.50	6.48	6.00	5.83	5.91
13	5.98	5.91	5.64	5.49	5.43	5.33	5.22	5.03	4.74	4.55	4.68	4.53
14	4.67	4.81	4.98	5.00	4.94	4.84	4.76	4.67	4.51	4.42	4.53	4.70
15	4.75	4.90	5.06	4.99	4.96	5.03	5.22	5.47	5.45	5.48	5.57	6.33
16	6.67	6.52	6.60	6.78	6.79	6.83	6.91	6.93	6.65	6.53	6.50	6.69
17	6.58	6.42	6.79	6.82	6.76	6.88	7.22	7.41	7.27	7.03	7.09	7.18
18	6.69	6.50	6.46	6.35	6.31	6.41	6.60	6.57	6.59	6.80	7.16	7.51
19	7.52	7.40	7.48	7.42	7.53	7.75	7.80	7.63	7.51	7.49	7.64	7.92
20	8.10	8.18	8.52	8.56	9.00	9.34	9.04	9.08	9.14	8.99	8.96	8.86
21	8.79	8.62	8.29	8.05	8.00	7.89	7.48	7.31	7.42	7.51	7.71	7.99

The first step is, as always, to plot the data, and this has been done in Figure D.1. The graph shows some interesting features. First, we note that there is no discernible seasonal variation. Secondly, we note a marked trend in the data from about 2% at the beginning of the series to over 7% at the end of the series. However, the trend is by no means regular and it would be a gross oversimplification to simply fit a straight line to the data. Thus the use of a trend-and-seasonal model, as in the Holt-Winters forecasting procedure, or the use of linear regression seems inappropriate. With a single important time series like this it is worth investing some effort on the analysis, and the use of the Box-Jenkins procedure is indicated.

The given time series is clearly non-stationary and some sort of differencing is required. To confirm this, the autocorrelation function of the data was calaculated and, as expected, the coefficients at low lags were all 'large' and positive as in Figure 2.3. The values at lags 1 to 5 were 0.97, 0.94, 0.92, 0.89 and 0.85, while the value at lag 24 was still as high as 0.34.

The simplest type of differencing procedure is to take first differences of the data, and this was done next. The autocorrelation function of the first differences turned out to be of a particularly simple form. The coefficient at lag 1 was 0.31, which is significantly different from zero. Here, we use the rule that values outside the range $\pm 2/\sqrt{N}$ are significantly different from zero,

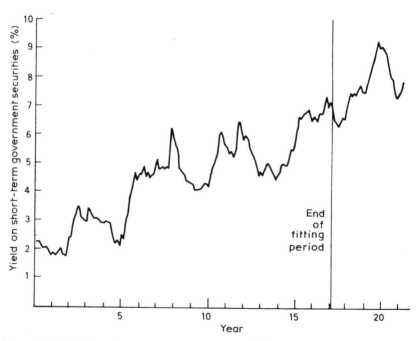

Figure D.1   Yield on short-term government securities.

where N is the number of terms in the differenced series. However, all subsequent coefficients up to about lag 20 were **not** significantly different from zero.

This tells us two things. First, the differenced series is now stationary and so no further differencing is required. Secondly, the fact that there is only one significant coefficient at lag 1 tells us that a very simple ARIMA model can be fitted. The autocorrelation function of the first-order moving average model (see Section 3.4.3) has the same form as the observed correlogram, and we can therefore try fitting an ARIMA(0, 1, 1) model to the observed data (or equivalently an ARMA(0, 1) model to the first differences).

Using MINITAB this model was fitted giving

$$\nabla X_t = Z_t + 0.37 Z_{t-1}$$

A similar model was also tried with an added constant on the right-hand side of the above equation, in view of the trend in the series. The residual sum of squares was not in fact much smaller and the estimated constant was not significantly different from zero. This is perhaps somewhat surprising at first sight, but variation due to the trend is 'small' compared with the variations about the trend.

We are now in a position to compute forecasts. The above model was actually fitted to the first 17 years' data with a view to comparing forecasts for the last four years with the observed values. However, the simple form of the fitted model means that the point forecasts from a given origin (e.g. the end of the 17th year) are all the same, although the confidence intervals do of course get wider as the distance ahead increases. For our fitted model, we wish to forecast $X_{N+1}, X_{N+2}, \ldots$ given data up to time N. Now $X_{N+1} = Z_{N+1} + 0.37 Z_N$. At time N, Z_{N+1} is unknown and is estimated as zero, while Z_N is estimated by the residual at time N. This cannot be written out explicitly as it involves the residual at time $N-1$ and is obtained iteratively. Nevertheless, we find that our forecast of X_{N+1} (and of $X_{N+2}, X_{N+3}, \ldots$) is given by $X_N + 0.37$(residual at time N). The difference between this forecast and X_{N+1} gives the residual at time $N+1$, which can in turn be used to compute forecasts from origin $N+1$.

It is interesting to note that the above forecasting procedure is in fact equivalent to exponential smoothing (see Section 5.2.2), which we have already noted is optimal for an ARIMA(0, 1, 1) model. As exponential smoothing is a special case of the Holt-Winters forecasting procedure with an appropriate choice of starting values, it turns out that the latter procedure could have been used after all. Nevertheless, the careful choice of an ARIMA model via the Box-Jenkins procedure is much more satisfying and reliable.

This example is one of the simplest Box-Jenkins analyses that the reader is likely to come across. An even simpler class of time series consists of those share price series where it is found that the first differences have a correlogram which cannot be distinguished from that of a completely random series. In other words none of the autocorrelation coefficients of the differenced series is significantly different from zero. This suggests that the (undifferenced) time series can be described by a random walk model.

In the next example, we present a much more difficult example of a Box-Jenkins analysis.

Example D.3   Airline passenger data   Table D.3 shows the monthly totals of international airline passengers from January 1949 to December 1960. This famous set of data has been analysed by many people including Box and Jenkins (1970, Chapter 9). The problem is to fit a suitable model to the data and produce forecasts for up to one year ahead.

As always we begin by plotting the data as in Figure D.2. This shows a clear seasonal pattern as well as an upward trend. The magnitude of the seasonal variation increases at the same sort of rate as the yearly mean levels, and this indicates that a **multiplicative** seasonal model is appropriate.

It is relatively straightforward to use the Holt-Winters forecasting procedure on these data. This procedure can cope with trend and with additive or multiplicative seasonality (see Section 5.2.3). The details of the analysis will not be given here as no particular difficulties arise and reasonable forecasts result. The reader who prefers a nice simple procedure is advised to use this method for these data.

Following Box and Jenkins (1970) and other authors, we will also try to fit an ARIMA model and consider the many problems which arise.

The first question which arises is whether or not to transform the data. The question of transformations is discussed in Section 2.4, and, in order to make

Table D.3   Monthly totals (X_t) of international airline passengers in thousands from January 1949 to December 1960

Year	Jan	Feb	Mar	Apr	May	June	July	Aug	Sept	Oct	Nov	Dec
1949	112	118	132	129	121	135	148	148	136	119	104	118
1950	115	126	141	135	125	149	170	170	158	133	114	140
1951	145	150	178	163	172	178	199	199	184	162	146	166
1952	171	180	193	181	183	218	230	242	209	191	172	194
1953	196	196	236	235	229	243	264	272	237	211	180	201
1954	204	188	235	227	234	264	302	293	259	229	203	229
1955	242	233	267	269	270	315	364	347	312	274	237	278
1956	284	277	317	313	318	374	413	405	355	306	271	306
1957	315	301	356	348	355	422	465	467	404	347	305	336
1958	340	318	362	348	363	435	491	505	404	359	310	337
1959	360	342	406	396	420	472	548	559	463	407	362	405
1960	417	391	419	461	472	535	622	606	508	461	390	432

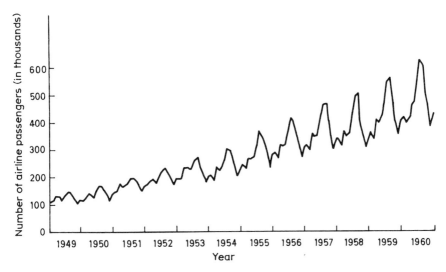

Figure D.2   Monthly totals of international airline passengers.

the seasonal effect additive, it looks as though we should take logarithms of the data. Box and Jenkins (1970, p. 303) also took logarithms of the data. To justify this, they simply say that 'logarithms are often taken before analysing sales data, because it is **percentage** fluctuation which might be expected to be comparable at different sales volumes.' This seems rather unconvincing, and the long discussion of Chatfield and Prothero (1973) indicates some of the problems involved in transformation. Nevertheless, it seems fairly clear that we should take logarithms of the data in this example, and so we shall proceed with the analysis on this assumption. It is easy to plot the logged data using MINITAB, which shows that the size of the seasonal variation is now roughly constant.

We now use MINITAB to plot the autocorrelation function (ac.f.) of the raw logged data and of various differenced series in order to see what form of differencing is required. The first 36 coefficients of the ac.f. of the logged data are all positive as a result of the obvious trend in the data. Some sort of differencing is clearly required. With monthly seasonal data, the obvious operators to try are ∇, ∇_{12}, $\nabla\nabla_{12}$ and ∇_{12}^2.

Letting $Y_t (= \log X_t)$ denote the logarithms of the observed data, we start by looking at the ac.f. of ∇Y_t. With N observations in the differenced series, a useful rule-of-thumb for deciding if an autocorrelation coefficient is significantly different from zero is to see if its modulus exceeds $2/\sqrt{N}$. Here the critical value is 0.17 and we find significant coefficients at lags 1, 4, 8, 11, 12, 13, 16, 20, 23, 24, 25 and so on. There is no sign that the ac.f. is damping out, and further differencing is required. For $\nabla_{12} Y_t$ the first nine coefficients are all positive and significantly different from zero, after which there is a long run of negative coefficients. The series is still non-stationary and so we next try $\nabla\nabla_{12}Y_t$. The first 26 coefficients are shown in Figure D.3 in the same form as presented by the MINITAB package. (One needs to look at more than two seasonal cycles of coefficients. I actually looked at the first 38 coefficients but only plot the first 26 to save space.) The number of terms in the differenced series is now $144-12-1=131$, and an approximate critical value is $2/\sqrt{131}=0.18$. We note 'significant' values at lags 1, 3, 12 and 23, but most of the other values are 'small' and there is no evidence of non-stationarity. Thus we choose to fit an ARIMA model to $\nabla\nabla_{12}Y_t$.

It has to be admitted that our choice of the differencing operator is somewhat subjective, and some authors prefer a more objective approach. One possibility is to choose the differencing operator so as to minimize the variance of the differenced series, but we did not try this here.

In order to identify a suitable ARMA model for $\nabla\nabla_{12}Y_t$, we need to calculate the partial ac.f. as well as the ac.f. (An alternative is to use the inverse ac.f. – see Section 11.4 – but this is not available in MINITAB.) The partial ac.f. is given in Figure D.4. As for the ac.f., coefficients whose moduli exceed 0.18 may, as a first approximation, be taken to be significantly different from zero. In the partial ac.f. we note significant values at lags 1, 3, 9 and 12. When 'significant'

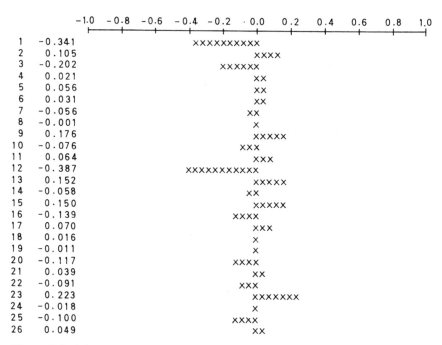

Figure D.3   The autocorrelation function of $\nabla\nabla_{12}\log_e X_t$ as printed by MINITAB.

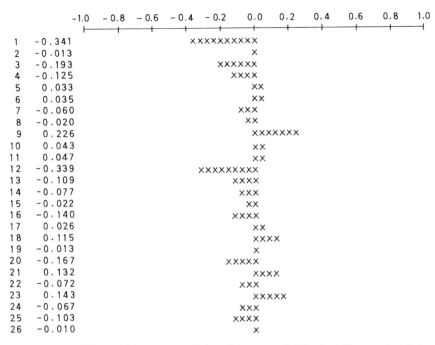

Figure D.4   The partial autocorrelation function of $\nabla\nabla_{12}\log_e X_t$ as printed by MINITAB.

values occur at unusual lagged values such as 9 they are usually ignored unless there is external information as to why such a lag should be important.

We should now be in a position to identify an appropriate seasonal ARIMA model to fit to the data. This means that we want to assess values of p, q, P and Q in the model defined by equation (4.16). The seasonal values P and Q are assessed by looking at the values of the ac.f. and partial ac.f. at lags 12, 24, 36, ... In this case the values are 'large' at lag 12 but 'small' at lags 24 and 36, indicating no autoregressive terms but one seasonal moving average term. Thus we take $P=0$ and $Q=1$. The values of the non-seasonal values p and q are assessed by looking at the first few values of the ac.f. and partial ac.f. The only 'significant' values are at lags 1 and 3, and these values are not easy to interpret. An AR(3) model, as suggested by the partial ac.f., would generally have a slowly decaying ac.f., while the MA(3) model, suggested by the ac.f., would generally have a slowly decaying partial ac.f. Noting that the coefficients at lag 3 are only just 'significant', we could, as a first try, take just one moving average term and set $p=0$ and $q=1$. If we now work out the standard error of the autocorrelation coefficient at lag 3 using a more exact formula, we find that it is not in fact significantly different from zero. This more exact formula (Box and Jenkins, 1970, p. 315) assumes that an MA(1) model is appropriate rather than a completely random series for which the formula $1/\sqrt{N}$ is appropriate. This result gives us more reliance on the choice of $p=0$ and $q=1$. Thus we now use MINITAB to fit a seasonal ARIMA model with $p=0$, $d=1$, $q=1$ and $P=0$, $D=1$, $Q=1$.

Setting $W_t = \nabla\nabla_{12} \log X_t$, the fitted model turns out to be

$$W_t = (1 - 0.396B)(1 - 0.614B^{12})a_t$$

MINITAB readily provides estimates of the residual variation as well as a variety of diagnostic checks. For example one should inspect the ac.f. of the residuals to see that 'none' of the coefficients is significantly different from zero (but remember that about 1 in 20 coefficients will be 'significant' at the 5% level under the null hypothesis that the residuals are random). One can also plot the observed values in the time series against the one-step-ahead forecasts of them. In this example there is no evidence that our fitted model is inadequate and so no alternative models will be tried.

MINITAB will readily provide forecasts of the logged series for up to 12 months ahead. The first three forecasts are 6.110, 6.056 and 6.178. These values need to be anti-logged to get forecasts for the original series.

I hope that this example has given some indication of the type of problem which may arise in a Box-Jenkins ARIMA analysis and the sort of subjective judgement which has to be made. A critical assessment of the method has been made in Section 5.4. The method sometimes works well, as in Example D.2. However, for the airline data, the variation is dominated by trend and seasonality and yet the complicated ARMA modelling is carried out on the differenced data from which most of the trend and seasonality have been removed. The reader must decide for himself if he thinks the method has been a success here. In my view, a trend-and-seasonal modelling approach (such as Holt-Winters) would not only be easier to apply but would arguably give more understanding of this particular set of data.

For the reader's assistance, I now give the MINITAB commands used in the above analysis. Numbers are manipulated by MINITAB in columns labelled from c1 to c50. We read the data in c1, and put the logs in c3 and the differenced series into c6, c7 and c8. I hope that the rest of the commands are self-explanatory. The commands are:

```
set c1
(read in data in rows of six observations each separated by one space followed
    by END or read from a file by e.g. read 'airline' c1)
tsplot c1
let c3 = log (c1)
tsplot c3
acf 38 c3
difference lag 1 c3 c6        (or just diff 1 c3 c6)
```

```
acf 38 c6
diff 12 c3 c7
acf 38 c7
diff lag 1 c7 c8
acf 38 c8
pacf 38 c8
arima (p=0 d=1 q=1) (P=0 D=1 Q=1) s=12 c3 res in c10 pred c11
acf 38 c10
plot c3 c11
arima (0 1 1) (0 1 1) s=12 c3;
forecast 12 c12.
stop
```

CONCLUDING REMARKS

I hope that the examples have illustrated some of the practical difficulties which arise in time-series analysis, particularly in regard to transformations, differencing and the subjective interpretation of correlograms and spectra. I also hope that the examples have demonstrated how the choice of analysis procedure depends on the objectives, which in turn depend on the particular practical situation. Clearly a careful statement of objectives is vital.

One important point which was not discussed in the examples is the possible use of ancillary information. In Example D.2, for instance, we have constructed a univariate model for the yield on short-term government securities which takes no account of other economic variables. As discussed in Chapter 5, it is not always possible to incorporate such ancillary information into a formal model but neither should the information be ignored. It may for example be advisable to adjust the forecasts subjectively if one knows that a change in bank rate or an oil crisis is imminent. This is just one example of how one should always be prepared to use one's common sense in carrying out a time-series analysis, or indeed any statistical investigation.

References

Abraham, B. and Ledolter, J. (1983) *Statistical Methods for Forecasting*, New York: Wiley.

Abraham, B. and Ledolter, J. (1986) Forecast functions implied by autoregressive integrated moving average models and other related forecast procedures. *Int. Statist. Rev.*, **54**, 51–66.

Akaike, H. (1967) Some problems in the application of the cross-spectral method. In *Spectral Analysis of Time Series* (ed. B. Harris), New York: Wiley.

Akaike, H. (1968) On the use of a linear model for the identification of feedback systems. *Ann. Inst. Statist. Math.*, **20**, 425–39.

Akaike, H. (1973) Contribution to the discussion of Chatfield and Prothero (1973). *J. R. Statist. Soc. A*, **136**, 330.

Akaike, H. (1978) Time-series analysis and control through parametric models. In *Applied Time Series Analysis* (ed. D. F. Findley), New York: Academic Press.

Anderson, B. D. O. and Moore, J. B. (1979) *Optimal Filtering*, Englewood Cliffs, NJ: Prentice Hall.

Anderson, T. W. (1971) *The Statistical Analysis of Time Series*, New York: Wiley.

Aoki, M. (1987) *State Space Modelling of Time Series*, Berlin: Springer-Verlag.

Armstrong, J. S. (1985) *Long-Range Forecasting* (2nd edn), New York: Wiley.

Ashley, R. (1988) On the relative worth of recent macroeconomic forecasts. *Int. J. Forecasting*, **4**, 363–76.

Astrom, K. J. (1970) *Introduction to Stochastic Control Theory*, New York: Academic Press.

Astrom, K. J. and Bohlin, T. (1966) Numerical identification of linear dynamic systems from normal operating records. In *Theory of Self-Adaptive Control Systems* (ed. P. M. Hammond), New York: Plenum Press, 94–111.

Astrom, K. J. and Eykhoff, P. (1971) System identification – a survey. *Automatica*, **7**, 123–62.

Barksdale, H. C. and Guffey, H. J. Jr (1972) An illustration of cross-spectral analysis in marketing. *J. Marketing Research*, **9**, 271–8.

Bartlett, M. S. (1966) *Stochastic Processes* (2nd edn), Cambridge: Cambridge University Press.

Bendat, J. S. and Piersol, A. G. (1986) *Random Data: Analysis and Measurement Procedures* (revised 2nd edn), New York: Wiley.

Besag, J. (1974) Spatial interaction and the statistical analysis of lattice systems. *J. R. Statist. Soc. B*, **36**, 192–236.

Bhattacharyya, M. N. (1974) Forecasting the demand for telephones in Australia. *Appl. Statist.*, **23**, 1–10.

Blake, I. F. and Lindsey, W. C. (1973) Level-crossing problems for random processes. *IEEE Trans.*, **IT-19**, no. 3, 295–315.

Bloomfield, P. (1976) *Fourier Analysis of Time Series: An Introduction*, New York: Wiley.

Bolstad, W. M. (1986) Harrison-Stevens forecasting and the multiprocess dynamic linear model. *Amer. Statist.*, **40**, 129–35.

Box, G. E. P. and Jenkins, G. M. (1968) Some recent advances in forecasting and control. Part I. *Appl. Statist.*, **17**, 91–109.

Box, G. E. P. and Jenkins, G. M. (1970) *Time Series Analysis, Forecasting and Control*, San Francisco: Holden-Day (revised edn published 1976).

Box, G. E. P. and MacGregor, J. F. (1974) The analysis of closed-loop dynamic-stochastic systems. *Technometrics*, **16**, 391–8.

Box, G. E. P. and Newbold, P. (1971) Some comments on a paper of Coen, Gomme and Kendall. *J. R. Statist. Soc. A*, **134**, 229–40.

Box, G. E. P. and Pierce, D. A. (1970) Distribution of residual auto-correlations in autoregressive-integrated moving average time-series models. *J. Amer. Statist. Ass.*, **65**, 1509–26.

Box, G. E. P. and Tiao, G. C. (1975) Intervention analysis with applications to economic and environmental problems. *J. Amer. Statist. Ass.*, **70**, 70–9.

Brass, W. (1974) Perspective in population prediction: illustrated by the statistics of England and Wales, *J. R. Statist. Soc. A*, **137**, 532–83.

Brillinger, D. R. (1975) *Time Series: Data Analysis and Theory*, New York: Holt, Rinehart and Winston (expanded edn published 1981).

Brillinger, D. R. and Krishnaiah, P. R. (eds) (1983) *Handbook of Statistics*, vol. 3, *Time Series in the Frequency Domain*, Amsterdam: North-Holland.

Brillinger, D. R. and Tiao, G. C. (eds) (1980) *Directions in Time Series*, Institute of Mathematical Statistics.

Brown, R. G. (1963) *Smoothing, Forecasting and Prediction*, Englewood Cliffs, NJ: Prentice-Hall.

Button, K. J. (1974) A critical review of techniques used for forecasting car ownership in discrete areas. *The Statistician*, **23**, 117–28.

Chang, I., Tiao, G. C. and Chen, C. (1988) Estimation of time-series parameters in the presence of outliers. *Technometrics*, **30**, 193–204.

Chatfield, C. (1974) Some comments on spectral analysis in marketing. *J. Marketing Research*, **11**, 97–101.

Chatfield, C. (1977) Some recent developments in time-series analysis. *J. R. Statist. Soc. A*, **140**, 492–510.

Chatfield, C. (1978a) The Holt-Winters forecasting procedure. *Appl. Statist.*, **27**, 264–79.

Chatfield, C. (1978b) Adaptive filtering: a critical assessment. *J. Opl. Res. Soc.*, **29**, 891–6.

Chatfield, C. (1979) Inverse autocorrelations. *J. R. Statist. Soc. A*, **142**, 363–77.

Chatfield, C. (1988a) What is the best method of forecasting? *J. Appl. Statist.*, **15**, 19–38.

Chatfield, C. (1988b) *Problem-Solving: A Statistician's Guide*, London: Chapman and Hall.

Chatfield, C. and Collins, A. J. (1980) *Introduction to Multivariate Analysis*, London: Chapman and Hall.

Chatfield, C. and Pepper, M. P. G. (1971) Time-series analysis: an example from geophysical data. *Appl. Statist.*, **20**, 217–38.

Chatfield, C. and Prothero, D. L. (1973) Box-Jenkins seasonal forecasting: problems in a case-study. *J. R. Statist. Soc. A*, **136**, 295–336.

Chatfield, C. and Yar, M. (1988) Holt-Winters forecasting: some practical issues. *The Statistician*, **37**, 129–40.

Cleveland, W. S. (1983) Seasonal and calendar adjustment. In *Handbook of Statistics*, vol. 3 (eds D. R. Brillinger and P. R. Krishnaiah), Amsterdam: North-Holland, 39–72.

Cleveland, W. S. and Devlin, S. J. (1982) Calendar effects in monthly time series: measurement and adjustment. *J. Amer. Statist. Ass.*, **77**, 520–8.

Cliff, A. D. and Ord, J. K. (1975) Model building and the analysis of spatial pattern in human geography (with discussion). *J. R. Statist. Soc. B*, **37**, 297–348.

Coen, P. J., Gomme, E. J. and Kendall, M. G. (1969) Lagged relationships in economic forecasting. *J. R. Statist. Soc. A*, **132**, 133–63.

Cooley, J. W., Lewis, P. A. W. and Welch, P. D. (1967) Historical notes on the fast Fourier transform. *IEEE Trans.*, **AU-15**, no. 2, 76–9.

Cox, D. R. (1981) Statistical analysis of time series. *Scand. J. Statist.*, **8**, 93–115.

Cox, D. R. and Isham, V. (1980) *Point Processes*, London: Chapman and Hall.

Cox, D. R. and Lewis, P. A. W. (1966) *The Statistical Analysis of Series of Events*, London: Chapman and Hall.

Cox, D. R. and Miller, H. D. (1968) *The Theory of Stochastic Processes*, London: Chapman and Hall.

Craddock, J. M. (1965) The analysis of meteorological time series for use in forecasting. *The Statistician*, **15**, 169–90.

Cramer, H. and Leadbetter, M. R. (1967) *Stationary and Related Stochastic Processes*, New York: Wiley.

Davies, N. and Newbold, P. (1979) Some power studies of a portmanteau test of time series model specification. *Biometrika*, **66**, 153–5.

Davis, M. H. A. and Vinter, R. B. (1985) *Stochastic Modelling and Control*, London: Chapman and Hall.

Durbin, J. and Murphy, M. J. (1975) Seasonal adjustment based on a mixed additive-multiplicative model. *J. R. Statist. Soc. A*, **138**, 385–410.

Eykhoff, P. (1974) *System Identification*, New York: Wiley Interscience.

Fildes, R. (1983) An evaluation of Bayesian forecasting. *J. Forecasting*, **2**, 137–50.

Fildes, R. (1985) Quantitative forecasting: the state of the art. Econometric models. *J. Op. Res. Soc.*, **36**, 549–80.

Fuller, A. T. (1970) *Non-linear Stochastic Control Systems*, Taylor and Francis.

Fuller, W. A. (1976) *Introduction to Statistical Time Series*, New York: Wiley.

Gardner, E. S. Jr (1983) Automatic monitoring of forecast errors. *J. Forecasting*, **2**, 1–21.

Gardner, E. S. Jr (1985) Exponential smoothing: the state of the art. *J. Forecasting*, **4**, 1–28.

Gardner, E. S. Jr (1988) A simple method of computing prediction intervals for time-series forecasts. *Man. Sci.*, **34**, 541–6.

Gardner, E. S. Jr and McKenzie, E. (1985) Forecasting trends in time series. *Man. Sci.*, **31**, 1237–46.

Gilchrist, W. (1976) *Statistical Forecasting*, London: Wiley.

Goodman, N. R., Katz, S., Kramer, B. H. and Kuo, M. T. (1961) Frequency response from stationary noise. *Technometrics*, **3**, 245–68.

Gooijer, J. G. de, Abraham, B., Gould, A. and Robinson, L. (1985) Methods for determining the order of an autoregressive-moving average process: a survey. *Int. Statist. Rev.*, **53**, 301–29.

Gottman, J. M. (1981) *Time Series Analysis*, Cambridge: Cambridge University Press.

Granger, C. W. J. (1966) The typical shape of an econometric variable. *Econometrica*, **34**, 150–61.

Granger, C. W. J. and Hatanaka, M. (1964) *Spectral Analysis of Economic Time Series*, Princeton: Princeton University Press.

Granger, C. W. J. and Hughes, A. O. (1968) Spectral analysis of short series – a simulation study. *J. R. Statist. Soc. A*, **131**, 83–99.

Granger, C. W. J. and Hughes, A. O. (1971) A new look at some old data: the Beveridge wheat price series. *J. R. Statist. Soc. A*, **134**, 413–28.

Granger, C. W. J. and Newbold, P. (1974) Spurious regressions in econometrics, *J. Econometrics*, **2**, 111–20.

Granger, C. W. J. and Newbold, P. (1986) *Forecasting Economic Time Series* (2nd edn), New York: Academic Press.

Griffiths, L. J. and Prieto-Diaz, R. (1977) Spectral analysis of natural seismic events using autoregressive techniques. *IEEE Trans.*, **GE-15**, no. 1, 13–25.

Grimmett, G. R. and Stirzaker, D. R. (1982) *Probability and Random Processes*, Oxford: Clarendon Press.

Gudmundsson, G. (1971) Time-series analysis of imports, exports and other economic variables. *J. R. Statist. Soc. A*, **134**, 383–412.

Gustavsson, I., Ljung, L. and Soderstrom, T. (1977) Identification of process in closed loop – identifiability and accuracy aspects. *Automatica*, **13**, 59–75.

Hamon, B. V. and Hannan, E. J. (1974) Spectral estimation of time delay for dispersive and non-dispersive systems. *Appl. Statist.*, **23**, 134–42.

Hannan, E. J. (1970) *Multiple Time Series*, New York: Wiley.

Hannan, E. J., Krishnaiah, P. R. and Rao, M. M. (eds) (1985) *Handbook of Statistics*, vol. 5: *Time Series in the Time Domain*, Amsterdam: North-Holland.

Harris, B. (ed.) (1967) *Spectral Analysis of Time Series*, New York: Wiley.

Harris, C. J. (1976) Problems in system identification and control. *Bull. IMA*, **12**, 139–50.

Harrison, P. J. (1965) Short-term sales forecasting. *Appl. Statist.*, **14**, 102–39.

Harrison, P. J. and Pearce, S. F. (1972) The use of trend curves as an aid to market forecasting. *Industrial Marketing Management*, **2**, 149–70.

Harrison, P. J. and Stevens, C. F. (1976) Bayesian forecasting (with discussion). *J. R. Statist. Soc. B*, **38**, 205–47.

Harvey, A. C. (1981a) *Time Series Models*. Oxford: Philip Allan.

Harvey, A. C. (1981b) *The Econometric Analysis of Time Series*, Oxford: Philip Allan.

Harvey, A. C. (1984) A unified view of statistical forecasting procedures. *J. Forecasting*, **3**, 245–75.

Harvey, A. C. (1989) *Forecasting, Structural Time Series Models and the Kalman Filter*, Cambridge: Cambridge University Press.

Hasan, T. (1983) Complex demodulation: some theory and applications. In *Handbook of Statistics*, vol. 3 (eds D. R. Brillinger and P. R. Krishnaiah), Amsterdam: North-Holland, 125–56.

Haubrich, R. A. and Mackenzie, G. S. (1965) Earth noise, 5 to 500 millicycles per second. *J. Geophysical Research*, **70**, no. 6, 1429–40.

Haugh, L. D. and Box, G. E. P. (1977) Identification of dynamic regression (distributed lag) models connecting two time series. *J. Amer. Statist. Ass.*, **72**, 121–30.

Hause, J. C. (1971) Spectral analysis and the detection of lead–lag relations, *Amer. Econ. Rev.*, **61**, 213–17.

Hillmer, S. (1984) Monitoring and adjusting forecasts in the presence of additive outliers. *J. Forecasting*, **3**, 205–15.

Hsu, H. P. (1967) *Fourier Analysis*, New York: Simon and Schuster.

Huot, G., Chiu, K. and Higginson, J. (1986) Analysis of revisions in the seasonal adjustment of data using X-11-ARIMA model-based filters. *Int. J. Forecasting*, **2**, 217–29.

Jacobs, O. L. R. (1974) *Introduction to Control Theory*, Oxford: Clarendon Press.

Jenkins, G. M. (1979) *Practical Experiences with Modelling and Forecasting Time Series*, Jersey: Gwilym Jenkins and Partners (Overseas) Ltd.

Jenkins, G. M. and McLeod, G. (1982) *Case Studies in Time Series Analysis*, Vol. 1, Lancaster: Gwilym Jenkins and Partners Ltd.

Jenkins, G. M. and Watts, D. G. (1968) *Spectral Analysis and its Applications*, San Francisco: Holden-Day.

Jones, R. H. (1965) A reappraisal of the periodogram in spectral analysis. *Technometrics*, **7**, 531–42.

Jones, R. H. (1974) Identification and autoregressive spectrum estimation. *IEEE Trans.*, **AC-19**, 894–7.

Jones, R. H. (1985) Time series analysis with unequally spaced data. In *Handbook of Statistics*, Vol. 5 (eds. E. J. Hannan *et al.*), Amsterdam: North-Holland, pp. 157–77.

Kendall, M. G., Stuart, A. and Ord, J. K. (1983) *The Advanced Theory of Statistics*, vol. 3 (4th edn), London: Griffin.

Kenny, P. B. and Durbin, J. (1982) Local trend estimation and seasonal adjustment of economic and social time series (with discussion). *J. R. Statist. Soc. A*, **145**, 1–41.

Kitagawa, G. (1987) Non-Gaussian state-space modelling of nonstationary time series (with discussion). *J. Amer. Statist. Ass.*, **82**, 1032–63.

Kohn, R. and Ansley, C. F. (1986) Estimation, prediction and interpolation for ARIMA models with missing data. *J. Amer. Statist. Ass.*, **81**, 751–61.

Koopmans, L. H. (1974) *The Spectral Analysis of Time Series*, New York: Academic Press.

Landers, T. F. and Lacoss, R. T. (1977) Some geophysical applications of autoregressive spectral estimates. *IEEE Trans.*, **GE-15**, no. 1, 26–32.

Levenbach, H. and Reuter, B. E. (1976) Forecasting trending time series with relative growth rate models. *Technometrics*, **18**, 261–72.

Ljung, G. M. and Box, G. E. P. (1978) On a measure of lack of fit in time series models. *Biometrika*, **65**, 297–303.

Ljung, L. (1987) *System Identification: Theory for the User*. Englewood Cliffs, NJ; Prentice-Hall.

Mackay, D. B. (1973) A spectral analysis of the frequency of supermarket visits. *J. Marketing Research*, **10**, 84–90.

Makridakis, S. (1986) The art and science of forecasting. *Int. J. Forecasting*, **2**, 15–39.

Makridakis, S., Andersen, A., Carbone, R., Fildes, R., Hibon, M., Lewandowski, R., Newton, J., Parzen, E. and Winkler, R. (1984) *The Forecasting Accuracy of Major Time Series Methods*, New York: Wiley.

Makridakis, S. and Hibon, M. (1979) Accuracy of forecasting: an empirical investigation (with discussion). *J. R. Statist. Soc. A*, **142**, 97–145.

Makridakis, S. and Wheelwright, S. C. (1977) Adaptive filtering: an integrated autoregressive/moving average filter for time series forecasting. *Opl. Res. Q*, **28**, 425–37.

Martin, D. R. (1983) Robust-resistant spectral analysis. In *Handbook of Statistics*, vol. 3 (eds D. R. Brillinger and P. R. Krishnaiah), Amsterdam: North-Holland, 185–219.

Maybeck, P. S. (1979) *Stochastic Models, Estimation and Control*, vol. 1, New York: Academic Press.

Meinhold, R. J. and Singpurwalla, N. D. (1983) Understanding the Kalman filter. *Amer. Statist.*, **32**, 123–7.

Montgomery, D. C. and Johnson, L. A. (1976) *Forecasting and Time Series Analysis*, New York: McGraw-Hill.

Mooers, C. N. K. and Smith, R. L. (1968) Continental shelf waves off Oregon. *J. Geophysical Research*, **73**, no. 2, 549–57.

Naylor, T. H., Seaks, T. G. and Wichern, D. W. (1972) Box-Jenkins methods: an alternative to econometric models. *Int. Statist. Rev.*, **40**, 123–37.

Neave, H. R. (1972a) Observations on 'Spectral analysis of short series – a simulation study' by Granger and Hughes. *J. R. Statist. Soc. A*, **135**, 393–405.

Neave, H. R. (1972b) A comparison of lag window generators. *J. Amer. Statist. Ass.*, **67**, 152–8.

Nelson, H. L. and Granger, C. W. J. (1979) Experience with using the Box-Cox transformation when forecasting economic time series. *J. Econometrics*, **10**, 57–69.

Newbold, P. (1981, 1984, 1988) Some recent developments in time-series analysis, I, II and III. *Int. Statist. Rev.*, **49**, 53–66; **52**, 183–92; **56**, 17–29.

Newbold, P. and Granger, C. W. J. (1974) Experience with forecasting univariate time-series and the combination of forecasts. *J. R. Statist. Soc. A*, **137**, 131–65.

Nicholson, H. (ed.) (1980) *Modelling of Dynamical Systems*, vols I and II, Stevenage, UK: Peter Peregrinus.

Otomo, T., Nakagawa, T. and Akaike, H. (1972) Statistical approach to computer control of cement rotary kilns. *Automatica*, **8**, 35–48.

Pagano, M. (1972) An algorithm for fitting autoregressive schemes. *Appl. Statist.*, **21**, 274–81.

Papoulis, A. (1984) *Probability, Random Variables, and Stochastic Processes* (2nd edn), New York: McGraw-Hill.

Parzen, E. (1962) *Stochastic Processes*, San Francisco: Holden-Day.

Parzen, E. (1982) ARARMA models for time-series analysis and forecasting. *J. Forecasting*, **1**, 67–82.

Parzen, E. (ed.) (1984) *Time Series Analysis of Irregularly Observed Data*, Proceedings of a symposium at Texas A and M University, New York: Springer.

Pena, D. and Box, G. E. P. (1987) Identifying a simplifying structure in time series. *J. Amer. Statist. Ass.*, **82**, 836–43.

Pierce, D. A. (1977) Relationships – and the lack thereof between economic time series, with special reference to money and interest rates. *J. Amer. Statist. Ass.*, **72**, 11–26.

Pierce, D. A. (1980) Some recent developments in seasonal adjustment. In *Directions in Time Series* (eds D. R. Brillinger and G. C. Tiao), Institute of Mathematical Statistics.

Pocock, S J. (1974) Harmonic analysis applied to seasonal variations in sickness absence. *Appl. Statist.*, **23**, 103–20.

Priestley, M. B. (1981) *Spectral Analysis and Time Series*, vols 1 and 2, London: Academic Press.

Priestley, M. B. (1983) The frequency domain approach to the analysis of closed-loop systems. In *Handbook of Statistics*, vol. 3 (eds D. R. Brillinger and P. R. Krishnaiah), Amsterdam: North-Holland, 275–91.

Priestley, M. B. (1988) *Non-linear and Non-stationary Time Series Analysis*, London: Academic Press.

Quenouille, M. H. (1957) *The Analysis of Multiple Time Series*, London: Griffin.

Reid, D. J. (1975) A review of short-term projection techniques. In *Practical Aspects of Forecasting* (ed. H. A. Gordon), London: Op. Res. Soc., 8–25.

Ripley, B. D. (1981) *Spatial Statistics*, Chichester: Wiley.

Roberts, J. B. and Gaster, M. (1980) On the estimation of spectra from randomly sampled signals: a method of reducing variability. *Proc. R. Soc. A*, **371**, 235–58.

Schwarz, R. J. and Friedland, B. (1965) *Linear Systems*, New York: McGraw-Hill.

Sneddon, I. H. (1972) *The Use of Integral Transforms*, New York: McGraw-Hill.

Snodgrass, F. E., Groves, G. W., Hasselmann, K. F., Miller, G. R., Munk, W. H. and Powers, W. H. (1966) Propagation of ocean swell across the Pacific. *Phil. Trans. R. Soc. London A*, **259**, 431–97.

Taylor, P. F. and Thomas, M. E. (1982) Short-term forecasting: horses for courses. *J. Op. Res. Soc.*, **33**, 685–94.

Tee, L. H. and Wu, S. M. (1972) An application of stochastic and dynamic models for the control of a papermaking process. *Technometrics*, **14**, 481–96.

Tetley, H. (1946) *Actuarial Statistics*, vol. 1, Cambridge: Cambridge University Press.

Tiao, G. C. and Box, G. E. P. (1981) Modelling multiple time series with applications. *J. Amer. Statist. Ass.*, **76**, 802–16.

Tomasek, O. (1972) Statistical forecasting of telephone time-series. *ITU Telecomm. J.*, December, 1–7.

Tong, H. (1977) Some comments on the Canadian Lynx data. *J. R. Statist. Soc. A*, **140**, 432 6.

Tyssedal, J. S. and Tjostheim, D. (1988) An autoregressive model with suddenly changing parameters and an application to stock market prices. *Appl. Statist.*, **37**, 353–69.

Vandaele, W. (1983) *Applied Time Series and Box-Jenkins Models*, New York: Academic Press.

Wallis, K. F. (1982) Seasonal adjustment and revision of current data. *J. R. Statist. Soc. A*, **145**, 76–85.

Wegman, E. J. and Wright, I. W. (1983) Splines in statistics. *J. Amer. Statist. Ass.*, **78**, 351–65.

West, M., Harrison, P. J. and Mignon, H. S. (1985) Dynamic generalized linear models and Bayesian forecasting (with discussion). *J. Amer. Statist. Ass.*, **80**, 73–83.

Wetherill, G. B. (1977) *Sampling Inspection and Quality Control* (2nd edn), London: Chapman and Hall.

Whittle, P. (1963) *Prediction and Regulation*, London: EUP (2nd edn published in 1983 by University of Minnesota Press).

Wiener, N. (1949) *Extrapolation, Interpolation, and Smoothing of Stationary Time-Series*, Cambridge, Mass.: MIT Press (note that this book was reissued in 1966 with the rather misleading title of *Time Series*).

Wright, G. and Ayton, P. (1987) *Judgemental Forecasting*, Chichester: Wiley.

Yaglom, A. M. (1962) *An Introduction to the Theory of Stationary Random Functions*, Englewood Cliffs, NJ: Prentice-Hall.

Young, P. (1974) Recursive approaches to time series analysis. *Bull. IMA*, **10**, 209–24.

Answers to exercises

Chapter 2

2.1 (b) There are various ways of assessing the trend and seasonal effects. The simplest method is to calculate the four yearly averages in 1967, 1968, 1969 and 1970, and also the average sales in each of periods I, II, ..., XIII (i.e. row and column averages). The yearly averages estimate trend, and the differences between the period averages and the overall average estimate the seasonal effects. With such a small downward trend, this rather crude procedure may well be adequate. It has the advantage of being easy to understand and to compute. A more sophisticated approach is to calculate a 13-month simple moving average, moving along one period at a time. This will give trend values for each period from period 7 to period 46. Extrapolate (by eye) the trend values to the end of the series. Calculate the difference between each observation and the corresponding trend value. Hence estimate the average seasonal effects in periods I, ..., XIII.

2.2 This exercise is easy to do using MINITAB. We find $r_1 = -0.55$. If a calculator is used, note that

$$\sum_{t=1}^{N-1} (x_t - \bar{x})(x_{t+1} - \bar{x})$$

$$= \sum_{t=1}^{N-1} x_t x_{t+1} - \bar{x}\left(\sum_{t=1}^{N-1} x_t + \sum_{t=2}^{N} x_t\right) + (N-1)\bar{x}^2$$

For this particular set of data, $\bar{x}$ happens to be one, and it is easier to subtract one from each observation.

2.4 $\pm 2/\sqrt{N} = \pm 0.1$. Thus r_7 is just 'significant', but unless there is some physical explanation for an effect at lag 7, there is no real evidence of non-randomness, as one expects 1 in 20 values of r_k to be 'significant' when data are random.

2.5 (b) ∇_{12}^2.

Chapter 3

3.1
$$\rho(k)= \begin{cases} 1 & k=0 \\ 0.56/1.53 & k=\pm 1 \\ -0.2/1.53 & k=\pm 2 \\ 0 & \text{otherwise} \end{cases}$$

3.3 $\mathrm{Var}(X_t)$ is not finite. $Y_t = Z_t + (C-1)Z_{t-1}$.
$$\rho_Y(k)= \begin{cases} 1 & k=0 \\ (C-1)/[1+(C-1)^2] & k=\pm 1 \\ 0 & \text{otherwise} \end{cases}$$

3.4 $\rho(k)=0.7^{|k|}$ $k=0, \pm 1, \pm 2, \dots$
      Note that this does not depend on μ.

3.6   We must have $|[\lambda_1 \pm \sqrt{(\lambda_1^2 + 4\lambda_2)}]/2| < 1$. If $\lambda_1^2 + 4\lambda_2 > 0$, we can easily show $\lambda_1 + \lambda_2 < 1$, and $\lambda_1 - \lambda_2 > -1$. If $\lambda_1^2 + 4\lambda_2 < 0$, roots are complex and we find $\lambda_2 > -1$.

3.8 $\gamma_Y(k) = 2\gamma_X(k) - \gamma_X(k+1) - \gamma_X(k-1)$

$$\gamma_Y(k)= \begin{cases} 2-2\lambda & k=0 \\ -\lambda^{|k|-1}(1-\lambda)^2 & k=\pm 1, \pm 2, \dots \end{cases}$$

3.9   All three models are stationary and invertible. For model (a) we have
      $X_t = Z_t + 0.3\, Z_{t-1} + 0.3^2 Z_{t-2} + \cdots$

3.11 First evaluate $\gamma(k)$. Find
$$\gamma(0) = \sigma_Z^2(1 + \beta^2 + 2\alpha\beta)/(1-\alpha^2)$$
$$\gamma(1) = \sigma_Z^2(\alpha + \alpha\beta^2 + \alpha^2\beta + \beta)/(1-\alpha^2)$$
$$\gamma(k) = \alpha\gamma(k-1) \quad k=2, 3, \dots$$
Note that $E(X_t Z_t) = \sigma_Z^2$, and $E(X_t Z_{t-1}) = (\alpha+\beta)\sigma_Z^2$.

3.12 (a) $p = d = q = 1$.
      (b) Yes.
      (c) 0.7, 0.64, 0.628.
      (d) 0.7, 0.15, 0.075, 0.037.

3.13 $\rho(k) = \dfrac{45}{38}\left(\dfrac{3}{4}\right)^{|k|} - \dfrac{7}{38}\left(\dfrac{1}{4}\right)^{|k|}$ $k=0, \pm 1, \pm 2, \dots$

      The AR(3) process is non-stationary, as the equation $(1 - B - cB^2 + cB^3) = 0$ has a root on the unit circle, namely $B=1$.

3.14 $\mathrm{Cov}[Y\,e^{i\omega t},\ \bar{Y}\,e^{-i\omega(t+\tau)}] = e^{-i\omega\tau}\,\mathrm{Cov}[Y, \bar{Y}]$ does not depend on t.

Chapter 4

4.2 $\Sigma(x_t - \bar{x})(x_{t-1} - \bar{x}) = \hat{\alpha}_1 \Sigma(x_{t-1} - \bar{x})^2 + \cdots + \hat{\alpha}_p \Sigma(x_{t-p} - \bar{x})(x_{t-1} - \bar{x})$
      . . . . .
      $\Sigma(x_t - \bar{x})(x_{t-p} - \bar{x}) = \hat{\alpha}_1 \Sigma(x_{t-1} - \bar{x})(x_{t-p} - \bar{x}) + \cdots + \hat{\alpha}_p \Sigma(x_{t-p} - \bar{x})^2$

All summations are for $t=(p+1)$ to N. These equations are the same as the sample Yule-Walker equations except that the constant divisor $\Sigma(x_t-\bar{x})^2$ is omitted and that γ_k is estimated from $(N-p)$ instead of $(N-k)$ cross-product terms.

4.3 When fitting an AR(2) process, π_2 is the coefficient α_2. For such a process, the first two Yule-Walker equations are: $\rho(2)=\alpha_1\rho(1)+\alpha_2$; $\rho(1)=\alpha_1+\alpha_2\rho(-1)=\alpha_1+\alpha_2\rho(1)$. Solve for α_2 by eliminating α_1.

4.4 $$\pi_j = \begin{cases} \rho(1)=9/21 & j=1 \\ \alpha_2=2/9 & j=2 \\ 0 & j>0 \end{cases}$$

Note that α_2 satisfies the equality in Exercise 4.3.

4.5 Values outside $\pm 2/\sqrt{100}=\pm 0.2$ are 'significant', i.e. r_1 and r_2. A second-order MA process has an ac.f. of this type.

4.6 The ac.f. of X_t indicates non-stationarity in the mean. Take first differences. None of the autocorrelations for ∇X_t is significantly large (all are less than $\pm 2/\sqrt{60}$). So an ARIMA(0, 1, 0) or random walk is a possible model. Would like to see a time plot of the data and get any relevant external information.

Chapter 5

5.4 Suppose we denote the model by
$$(1-\alpha B^{12})W_t=(1+\theta B)Z_t \quad \text{or}$$
$$X_t=Z_t+\theta Z_{t-1}+(1+\alpha)X_{t-12}-\alpha X_{t-24}'$$
Then
$$\hat{x}(N, 1)=(1+\hat{\alpha})x_{N-11}-\hat{\alpha}x_{N-23}+\hat{\theta}Z_N \quad \text{and}$$
$$\hat{x}(N, k)=(1+\hat{\alpha})x_{N+k-12}-\hat{\alpha}x_{N+k-24} \quad k=2, 3, \ldots, 12$$

5.5 $\hat{x}(N, 1)=1.2x_N-0.2x_{N-1}$
$\hat{x}(N, 2)=1.2\hat{x}(N, 1)-0.2x_N$
$\text{Var}[e(N, k)]=\sigma_Z^2$ when $k=1$, $1.49\sigma_Z^2$ when $k=2$ and $1.90\sigma_Z^2$ when $k=3$.

Chapter 6

6.1 (a) $f(\omega)=\sigma_Z^2/\pi(1-2\lambda \cos \omega+\lambda^2)$ for $0<\omega<\pi$
(b) $f^*(\omega)=(1-\lambda^2)/\pi(1-2\lambda \cos \omega+\lambda^2)$

6.2 (a) $f(\omega)=\sigma_Z^2[3+2(2 \cos \omega+\cos 2\omega)]/\pi$
(b) $f(\omega)=\sigma_Z^2[1.34+2(0.35 \cos \omega-0.3 \cos 2\omega)]/\pi$

6.3 The non-zero mean makes no difference to the acv.f., ac.f. or spectrum.
$$\gamma(k) = \begin{cases} 1.89 \sigma_Z^2 & k=0 \\ 1.2 \sigma_Z^2 & k=\pm 1 \\ 0.5 \sigma_Z^2 & k=\pm 2 \\ 0 & \text{otherwise} \end{cases}$$

$$\rho(k) = \begin{cases} 1 & k=0 \\ 1.2/1.89 & k=\pm 1 \\ 0.5/1.89 & k=\pm 2 \\ 0 & \text{otherwise} \end{cases}$$

6.4 $\rho(k) = \displaystyle\int_0^\pi f^*(\omega)\cos\omega k\ d\omega$

$$= -\left[\frac{2\cos\omega k}{\pi^2 k^2}\right]_0^\pi$$

6.5   Clearly $E[X(t)] = 0$

$E[X(t)X(t+u)] = P[X(t)$ and $X(t+u)$ have same sign$]$
$\qquad\qquad\qquad - P[X(t)$ and $X(t+u)$ have opposite sign$]$

Hint: P(even number of changes in time u)

$$= e^{-\lambda u}\left[1 + \frac{(\lambda u)^2}{2!} + \frac{(\lambda u)^4}{4!} + \cdots\right]$$

$$= e^{-\lambda u}(e^{\lambda u} + e^{-\lambda u})/2$$

Chapter 7

7.6 $2N \left/ \displaystyle\sum_{k=-M}^{+M} \frac{1}{4}\left(1 + \cos\frac{\pi k}{M}\right)^2\right. = 8N/(3M+2) \to 8N/3M$ as $M \to \infty$

Chapter 8

8.1 $f_{xy}(\omega) = \dfrac{\sigma_Z^2}{\pi}\,[\beta_{11}\beta_{21} + \beta_{12}\beta_{22} + \beta_{21}\,e^{-i\omega} + \beta_{12}\,e^{+i\omega}]$

8.2   Use the fact that $\mathrm{Var}[\lambda_1 X(t) + \lambda_2 Y(t+\tau)] \geqslant 0$ for any constants λ_1, λ_2 (e.g. put $\lambda_1 = \sigma_Y$, $\lambda_2 = \sigma_X \to \rho_{xy} \geqslant -1$).

$$\gamma_{xy}(k) = \begin{cases} 0.84\,\sigma_Z^2 & k=0 \\ -0.4\,\sigma_Z^2 & k=1 \\ +0.4\,\sigma_Z^2 & k=-1 \\ 0 & \text{otherwise} \end{cases}$$

$c(\omega) = 0.84\,\sigma_Z^2/\pi$
$q(\omega) = -0.8\,\sigma_Z^2 \sin\omega/\pi$

$\alpha_{xy}(\omega) = \dfrac{\sigma_Z^2}{\pi}\,\sqrt{(0.84^2 + 0.8^2 \sin^2\omega)}$

$\tan\phi_{xy}(\omega) = 0.8\sin\omega/0.84$
$C(\omega) = 1$
All for $0 < \omega < \pi$.

Chapter 9

9.1 Selected answers only:

(a)
$$h_k = \begin{cases} \frac{1}{2} & k = \pm 1 \\ 1 & k = 0 \\ 0 & \text{otherwise} \end{cases}$$

$$S_t = \begin{cases} 0 & t < -1 \\ \frac{1}{2} & t = -1 \\ 1\frac{1}{2} & t = 0 \\ 2 & t \geqslant 1 \end{cases}$$

$H(\omega) = \frac{1}{2} e^{-i\omega} + 1 + \frac{1}{2} e^{i\omega} = 1 + \cos \omega$
$G(\omega) = H(\omega)$ since $H(\omega)$ is real. $\phi(\omega) = 0$.

(b) $H(\omega) = \dfrac{1}{5} + \dfrac{2}{5} \cos \omega + \dfrac{2}{5} \cos 2\omega$

(c)
$$h_k = \begin{cases} +1 & k = 0 \\ -1 & k = +1 \\ 0 & \text{otherwise} \end{cases}$$

(a) and (b) are low-pass filters. (c) and (d) are high-pass filters. The combined filter has

$$H(\omega) = (1 + \cos \omega)\left(\frac{1}{5} + \frac{2}{5} \cos \omega + \frac{2}{5} \cos 2\omega\right)$$

9.2 (a) $H(\omega) = g\, e^{-i\omega\tau}$ $\qquad\qquad\qquad \omega > 0$
     (b) $H(\omega) = g(1 - i\omega T)/(1 + \omega^2 T^2)$ $\qquad \omega > 0$

9.4 $\gamma(k) = \begin{cases} \alpha^{|k/2|} \sigma_Z^2/(1 - \alpha^2) & k \text{ even} \\ 0 & k \text{ odd} \end{cases}$

$f_X(\omega) = f_Z(\omega)/(1 - 2\alpha \cos 2\omega + \alpha^2)$ $\qquad 0 < \omega < \pi$

Chapter 10

10.2 We have

$$X_2 = \mu_2 + n_2$$
$$X_1 = \mu_1 + n_1 = \mu_2 - \beta_1 + n_1$$
$$\qquad = \mu_2 - \beta_2 + n_1 + w_2$$

Thus

$$\begin{bmatrix} X_2 \\ X_1 \end{bmatrix} = \begin{bmatrix} 1 & 0 \\ 1 & -1 \end{bmatrix} \begin{bmatrix} \mu_2 \\ \beta_2 \end{bmatrix} + \begin{bmatrix} n_2 \\ n_1 + w_2 \end{bmatrix}$$

So

$$\hat{\theta}_2 = \begin{bmatrix} \hat{\mu}_2 \\ \hat{\beta}_2 \end{bmatrix} = \begin{bmatrix} 1 & 0 \\ 1 & -1 \end{bmatrix}^{-1} \begin{bmatrix} X_2 \\ X_1 \end{bmatrix} = \begin{bmatrix} 1 & 0 \\ 1 & -1 \end{bmatrix} \begin{bmatrix} X_2 \\ X_1 \end{bmatrix}$$

$$= \begin{bmatrix} X_2 \\ X_2 - X_1 \end{bmatrix}$$

(see equation (10.17))

To find P_2, var($\hat{\beta}_2$) for example is

$$E(X_2 - X_1 - \beta_2)^2 = E(n_2 - n_1 + \beta_2 - w_2 - \beta_2)^2 = 2\sigma_n^2 + \sigma_w^2$$

Using the Kalman filter, find

$$\hat{\theta}_{3|2} = \begin{bmatrix} 2X_2 - X_1 \\ X_2 - X_1 \end{bmatrix}, P_{3|2} = \sigma_n^2 \begin{bmatrix} 5 & 3 \\ 3 & 2 \end{bmatrix}, K_3 = \begin{bmatrix} 5/6 \\ 1/2 \end{bmatrix}$$

Hence $\hat{\theta}_3$ using equation (10.13).

10.3 (a) $h^T = [1, 0]$; $G = \begin{bmatrix} 0 & 1 \\ 0 & 0 \end{bmatrix}$; $w_t^T = [1, \beta]Z_t$. Alternatively if we take

$\theta_t^T = [X_t, Z_t]$, then $G = \begin{bmatrix} 0 & \beta \\ 0 & 0 \end{bmatrix}$.

(b) Try $\theta_t^T = [X_t, \hat{X}(t, 1), \hat{X}(t, 2)] = [X_t, \beta_1 Z_t + \beta_2 Z_{t-1}, \beta_2 Z_t]$ with

$$G = \begin{bmatrix} 0 & 1 & 0 \\ 0 & 0 & 1 \\ 0 & 0 & 0 \end{bmatrix}.$$

Or try $\theta_t^T = [X_t, Z_t, Z_{t-1}]$.

Author index

This index does not include entries in the reference section (pp. 222–8) where source references may be found. Where the text refers to an article or book by more than one author, only the first-named author is listed here.

Subject index